"*Aqua Shock* confronts an unfortunately little-known fact—the United States is in the midst of a water crisis. Drought and overuse are growing problems, proliferation of chemicals is causing deformities in fish and impacts to human health, and pollution from storm water runoff is damaging our aquatic systems. A talented journalist, Susan Marks highlights these and other challenges, as well as the competing and sometimes conflicting responses. This is not an issue that can be ignored. Americans need to read this book."

—KRISTINE STRATTON
Executive Director, Waterkeeper Alliance

"Susan J. Marks' book, *Aqua Shock: Water in Crisis*, brings to our attention the dire straits that America faces in the future because of water shortages and pollution of existing sources. It is time that the American public awoke to these conditions and took steps, however costly, to change them. *Aqua Shock* will help achieve this desired goal."

—JOEL A. TARR
Richard S. Caliguiri University Professor of History
and Policy, Carnegie Mellon University

"Across the United States and throughout the world, we have rivers so overallocated that they no longer reach the sea; waterways in our communities so severely degraded that we'd never dream of allowing our children to swim, drink, or fish even near them; and management practices around it all that are often entirely devoid of the science we've spent generations discovering. Our water is in crisis today even without the exacerbating pressures global climate change is already starting to unveil for tomorrow. 'Water' is truly *the* challenge of the twenty-first century—*our* century—and we have to start having a serious mainstream conversation about the legacy we are leaving our children and grandchildren. In *Aqua Shock: Water in Crisis*, Susan Marks does a great job of getting that conversation started."

—ALEXANDRA COUSTEAU
Explorer, Filmmaker, and Water Advocate
(www.AlexandraCousteau.org)

"The first step toward making the changes that need to be made to protect our water resources is raising public awareness, and with *Aqua Shock*, Susan J. Marks has delivered what is needed—a highly readable and tremendously informative account of what she quite rightly calls America's water crisis. Thankfully, Marks not only spells out in rich detail how and why we as a country find ourselves in the present predicament, but also how we can emerge from it and satisfy our thirst for water in a safe and sane manner. The days of profligate water use are over. They will never return nor should they. With the appropriate steps, we can ensure an adequate supply of clean water for all Americans. Future generations need not be saddled with this crisis. It is within our power to solve this problem, and we must."

—RONALD F. POLTAK
EPA Lifetime Achievement Award Winner (2006) and
Executive Director of the New England Interstate
Water Pollution Control Commission

"Water is a renewable but limited resource. In many communities, water resources are being depleted faster than they are replenished. If this rate is not reversed, freshwater can no longer be considered a renewable resource. We use and abuse water in ways that are not sustainable. Once a free gift of nature, water is now a costly commodity. We are throwing away the very elixir of life. In *Aqua Shock*, author Susan J. Marks gives a clear-headed summary of the diverse locations and predicaments of water shortages in the United States, along with steps we must take to right the balance."

—DONALD WATSON
Architect, Author, and Specialist in Urban Design and Sustainability

"*Aqua Shock* brings to the forefront a topic which has slowly been building to a crescendo for decades. As the world's population grew sixfold in the last hundred years, demand for water has outstripped abundant, clean supply. Susan J. Marks reveals, in an easy-to-read journalistic style, the water confrontations that are rapidly emerging and will erupt in the coming decades. *Aqua Shock* concludes with a series of positive actions that all Americans can take to reverse the fundamentals. The recommendations are given both prescriptively and by showing what other individuals, groups, and communities are doing to rise to the occasion. Kudos to Susan Marks for elevating water to a critical issue on a national level."

—TOM BINNINGS
Colorado-Based Water Policy and Economics Adviser
and Partner, Summit Economics

"The discussions presented in *Aqua Shock* are very timely considering the tough water-resources decisions being made across the country. With more than 49 inches of rainfall per year, Arkansas truly is a water-rich state. However, withdrawals from aquifers at a rate that is not sustainable have caused water-level declines of one foot per year or more and the development of large cones of depression in eastern and southern Arkansas."

—D. TODD FUGITT R.P.G.
Geology Supervisor, Arkansas
Natural Resources Commission

"The hazards of extended drought are not about loss of green lawns and water rationing, but rather social disruption, conflict, and mass migration of people to greener pastures. Our water-planning scenarios must include worst-case scenarios. Thanks to Susan Marks and her years of journalism experience in providing this summary of social conflicts over the sharing of water."

—KENNETH R. WRIGHT
Paleohydrologist and Founder of Denver-based
Wright Water Engineers

"Susan Marks' timely new book on the coming water crisis is aptly named. What *Future Shock* did for change, *Aqua Shock* does for water. Water is the axis, the yoke of the interdependent troika of energy, food, and high technology that will drive the new economy. We have long underinvested in our water infrastructure. In *Aqua Shock*, Marks brings her objective viewpoint and cheerful voice to the set of difficult, but not insurmountable, issues that define water today. Marks reminds us that water is indeed gold and that managing water as a first-priority resource is essential to a successful twenty-first-century civilization."

—ADAM W. GRAVLEY
Seattle-Based Attorney Specializing in Water Issues, GordonDerr LLP

"We must shift from business as usual—what we at Earth Policy Institute call Plan A—to a plan of action to save our environment. This new plan includes restoration of aquifers. We can arrest the fall in water tables by increasing water productivity—getting the most use from every drop of water that we do have. That means, as *Aqua Shock* details, recycling, reuse, conservation, and rethinking how and where we use water today. We can do it, but time is not on our side. We must act now before it's too late."

—LESTER R. BROWN
President of Earth Policy Institute and
Author of *Plan B 4.0: Mobilizing to Save Civilization*

"Water is not only essential to human life as the single most important input for both advanced and underdeveloped economies, it is also essential to our enduring prosperity. Competing demands for this scarce and mismanaged resource will shape global business and geopolitics in the twenty-first century. *Aqua Shock* and the message it delivers are key to understanding water's inescapable influence."

—JOHN I. DICKERSON
CEO, Summit Global Management

AQUA SHOCK

AQUA SHOCK

WATER IN CRISIS

REVISED AND UPDATED

SUSAN J. MARKS

WILEY

John Wiley & Sons, Inc.

For general information on our other products and services or for technical
support, please contact our Customer Care Department within the United States at
(800) 762-2974, outside the United States at (317) 572-3993 or fax (317) 572-4002.

Wiley also publishes its books in a variety of electronic formats. Some content that
appears in print may not be available in electronic books. For more information
about Wiley products, visit our web site at www.wiley.com.

ISBN 978-0-470-91806-7 (cloth); ISBN 978-0-470-88344-0 (ebk);
ISBN 978-0-470-88444-7 (ebk); ISBN 978-0-470-88554-4 (ebk)

Printed in the United States of America
10 9 8 7 6 5 4 3 2 1

MIX
Paper from
responsible sources
FSC® C005928

The ice is melting off the window . . .

Thank you, E&H;
and to John and Emma for your patience and understanding.

CONTENTS

ACKNOWLEDGMENTS

This book would not have been possible without the cooperation and efforts of many, many people who willingly shared their wisdom, expertise, and personal experiences. That includes scholars, scientists, educators, experts, lawyers, prognosticators, businesspeople, administrators, journalists, farmers, and friends as well as ordinary people with extraordinary tales to tell.

Throughout *Aqua Shock*, I have tried to treat the issues evenhandedly, but water is a controversial and contentious topic. There are as many sides to an issue as there are participants. The issues, disputes, and situations constantly change, too, especially as more people begin to wake up to the severity of the world's water crisis. I've relied a great deal on data collection and studies from various organizations, including two federal government agencies: the U.S. Geological Survey and the National Oceanic and Atmospheric Administration. These two agencies are key sources of data and scientific studies, and many private and public entities, as well as other government organizations, use their research as a basis for their own statistics and reports.

In addition, special thanks to the following people:

- Chris Gandomcar, for the long trail rides and conversations that introduced me to water.
- Stephen Isaacs, Bloomberg's Chicago-based Editor-at-Large, and Laura Walsh, a Hoboken, New Jersey–based Senior Editor at John Wiley & Sons, for understanding the importance of water, and believing in *Aqua Shock*.
- Cynthia Zigmund, my Chicago-based book agent from Second City Publishing Services, for the initial idea of a book "on water," and her unwavering support.
- Kris McGovern, my longtime friend, copy editor, and stalwart support.

- U.S. Geological Survey experts and staff, for their cooperation.
- Those individuals who so graciously took the time to provide their insightful testimonials and comments on *Aqua Shock* and the water crisis.
- Last, but not least, all the unsung heroes—those behind-the-scenes staffers at all the organizations I have written about. They're the people who help make it all happen.

SUSAN J. MARKS

AQUA SHOCK

INTRODUCTION

America is running out of water! Wherever you live, whatever the weather, whether water pours plentifully from your faucets, sooner rather than later your tap could run dry.

Tens of thousands of acres of the nation's farmland already are parched. Reservoirs, lakes, and streams across the country have dried up. Natural and manmade pollutants, poisons, and contaminants taint water resources. States, growing cities, businesses, neighbors, and one-time friends now fight over the right to take what's left from our fast-shrinking rivers and lakes, and from the underground water supplies known as *aquifers* that hold our freshwater.

Water has become the golden commodity of the twenty-first century. Once plentiful and pure, today it's become a finite resource like oil. The difference: There is no alternative. All living things need water to survive. Prices are rising, too, though not high enough or fast enough to make a difference for most Americans.

Some battles for this new millennium's "clear gold" are reminiscent of frontier-style, guns-drawn shootouts at the OK Corral; others end up as years-long, megamillion-dollar fights in courts, legislatures, or Congress. Whatever the battlefield, the fights are equally acrimonious and devastating to the losers.

In Arizona pecan farmers watch their trees and livelihood wither in a battle with an industrial neighbor, who they claim depleted the area's aquifer. After the aquifer's water level drops by half—from 32 feet to only 16 feet—the trees die of thirst.

The state of Mississippi sues the city of Memphis, Tennessee, and its Memphis Light, Gas and Water Company to the tune of $1 billion claiming Memphis pilfered tens of millions of gallons of water from Mississippi's aquifer. Memphis walks away the apparent winner—for the moment—but only after a years-long battle in which the lower courts rule in its favor, and then on appeal the

1

case goes all the way to the U.S. Supreme Court, which refuses to hear it.

In Colorado, the courts order more than 400 water wells shut down, plugged, leaving farmers and families high and dry. The courts determine these people "stole" water they didn't own the rights to, even though the wells are on their individual properties.

Residents of Chattanooga, Tennessee, may have to trade their Tennessee flags for that of Georgia if the two states can't settle their old-fashioned border war. The battle is precipitated by a controversial 1818 land survey with the prize, access to the Tennessee River and its billions of gallons of water.

In 2007, victory is bittersweet for tens of thousands of current and former U.S. Marines and their families when, after years of dead ends, denials, and verbal sparring, the government finally releases its analysis of a tainted water supply at Marine Corps Base Camp Lejeune, North Carolina. From 1957 to 1987, the drinking water at Tarawa Terrace family housing was contaminated with tetrachloroethylene (PCE), a carcinogen that can cause cancers and birth defects. The military shut down the housing units in 1987, but the report wasn't completed until 20 years later!

Water battles today take on many issues and rage coast to coast. Large-scale water disputes once were rare, and only arose in desert states or between frontier farmers and ranchers. But that was before huge populations, urban and rural sprawl, years of overbuilding and development, drought, climate change, pollution, and more took their toll. Clean water once was abundant, with plenty to go around. But that's not necessarily the case anymore, especially if you factor in the finite supply mixed with burgeoning demand, and growing water pollution—natural and otherwise. Earth's essential, no-longer-so-easily-renewable resource is in short supply.

Aqua Shock looks realistically at the water crisis in America. It touches on global issues and connections; explains where our water comes from, what's happening to it, and why; examines the poorly understood and highly complicated water laws and government edicts that control water supplies; discusses who does and doesn't own the rights to the water; describes how our groundwater is polluted and depleted; and considers what, if anything, can be done to ease the crisis.

This isn't another book filled with corporate-speak or grand-standing for a cause, and it doesn't dwell on the technicalities of the world's water or the shortcomings of conservation or development. Neither does it single out states, developers, groups, or individuals for ridicule or blame.

Instead, *Aqua Shock* is a simple description of our nation's water as a shrinking resource, and the problems, issues, and complexities associated with it. The book brings home the shocking realities of America's battle for water with real-life illustrations of the thirst and tribulations of individuals, companies, towns, cities, states, and regions. We turn to real stories from real people who give this global issue a human face in our own neighborhoods.

Before anyone shrugs off *Aqua Shock* as scare tactics, the woes of somewhere else, or more rhetoric from environmentalists, politicians, or corporations, keep in mind that our nation's midsection—with its withered fields and shrinking groundwater supplies—fulfills much of America's (and the world's) appetite for food, water, and—now with corn-based ethanol—fuel. That midsection stretches from North Dakota to Texas and from California to Nebraska. Sometimes heavy rains and flooding do little more than add to the crisis. They can deliver too much water at the wrong place at the wrong time without the infrastructure to harness it, and exacerbate water pollution issues in the process.

The U.S. water shortage isn't confined to the Great Plains or the West. At least 36 states across the country expect water shortages of some kind by 2013, and that's not even factoring in drought or changing climate conditions, according to a 2003 report from the U.S. General Accounting Office.[1] Forty-six states are expected to be under drought conditions by 2013.

If you think that it's not in your neighborhood, look more closely:

- The 5 million residents of Atlanta, Georgia, were shocked into reality in 2007 when their main water source, Lake Lanier, nearly dried up. By spring 2009, rains had eased the years-long drought and Atlantans' water worries, or so it seemed. In 2010, a federal judge drove another nail in the city's water coffin when he ruled Atlanta doesn't have the right to take water from Lake Lanier. Later that same year, drought returned.

- North Carolina had its driest winter in 113 years in 2007, according to data from the National Climatic Data Center, part of the U.S. National Oceanic and Atmospheric Administration. Rains in 2009 eased conditions in some parts of the state, though others continued to suffer. Summer 2010 brought record high temperatures to parts of the state, with drought persisting across southeastern portions.[2]
- New Jersey's Department of Environmental Protection issued a statewide drought watch in September 2010 after the state experienced its warmest summer (June–August) since weather data collection began in 1895, and its driest summer since 1966, according to State Climatologist David Robinson of Rutgers University. The drought watch was lifted for most of the state at the end of October.[3]
- Florida, a peninsula (meaning it's surrounded on three sides by water), averages more than 50 inches of rain a year, yet some areas regularly face water-shortage emergencies. With rainfall totals 70 percent below normal, by mid-March 2009, Tampa Bay Water's regional reservoir ran out of water. It has since filled back up as drought conditions eased, but the area's growing water demands are not shrinking, and neither is the population. In 2010, Orlando, Florida, had its driest summer since 1948.[4]
- Rain forests, paradise, and some of the wettest spots on earth aren't immune, either. In July 2010, the U.S. Department of Agriculture designated the Hawaiian counties of Hawaii, Maui, Honolulu, and Kauai disaster areas due to losses related to ongoing drought.

Water shortage is a national problem we no longer can ignore. It's global in scope, too. Here are some numbers:

- More than 1 billion people worldwide do not have access to minimal amounts of clean water, according to United Nations data.[5]
- In Latin America alone, approximately 76 million people lack safe water, according to the World Bank.

- Every year 1.8 million children die as a result of diarrhea and other diseases caused by unsafe water and poor sanitation, according to the United Nations report mentioned earlier.
- By 2035, as many as 3 billion people may live in areas with severe water shortages, especially if they live in Africa, the Middle East, or South Asia, as the World Bank predicts they will.

The issue for Americans isn't simply a result of population growth or water demand, drought, development, or pollution. It's all of that and more.

Aqua Shock begins with a look at our nation's water supply: where it is, what's happened to it, the global perspective, and why we should be worried. Then we examine why our water is in short supply: drought, development practices, population changes, overuse, regulation (or lack of it), worn-out sewer systems that leach away precious freshwater supplies, and contaminants—both natural and man-made. We'll also delve into the morass of rules and regulations that govern water: who owns it, who doesn't, and the "water gods" that control this precious resource. These "gods" are often little-known, extremely powerful individuals in many areas of the country who, by law and sometimes behind the scenes, play a big role in whether you, your neighbor, your neighbor's neighbor, or an entire town or city does or does not get water. We'll also look briefly at whether it's possible to save our water and how that can be accomplished.

After reading *Aqua Shock,* you'll better understand why our water is a finite resource, and how ordinary individuals can help turn the thirsty tide with the right information and direction.

Water is a broad issue and—through the lens of *Aqua Shock*—anything but dry, so let's get started.

NOTES

1. U.S. General Accounting Office, Report to Congressional Requesters, GAO-03-514, "Freshwater Supply: States' Views of How Federal Agencies Could Help Them Meet the Challenges of Expected Shortages," July 2003, 8, http://www.gao.gov/new.items/d03514.pdf.
2. NOAA Satellite and Information Service, "State of the Climate National Overview," August 2010, http://lwf.ncdc.noaa.gov/sotc/?report=national.

3. New Jersey Department of Environmental Protection, "DEP Institutes Statewide Drought Watch," press release, September 8, 2010, http://www .state.nj.us/dep/newsrel/2010/10_0088.htm; "Drought Watch Lifted for Most of New Jersey," press release, October 26, 2010, http://www.nj.gov/dep/ newsrel/2010/10_0117.htm.

4. Florida Department of Environmental Protection, "Florida Drought Conditions, Frequently Asked Questions," www.dep.state.fl.us/. Drought/faq .htm; Tampa Bay Water, "Tampa Bay Regional Water Supply and Drought Index," April 6, 2009, www.tampabaywater.org/newsarticles/152.aspx; NOAA Satellite and Information Service, "State of the Climate National Overview," August 2010, http://lwf.ncdc.noaa.gov/sotc/?report=national.

5. First United Nations World Water Development Report, "Water for People, Water for Life" (2003); Second United Nations World Water Development Report, "Water, a Shared Responsibility" (2006); United Nations Development Programme, "Human Development Report" (2006), http://www.unesco.org/ water/news/newsletter/183.shtml#know.

CHAPTER 1

LIQUID GOLD

We have really no idea how bad off we are. It's a disaster waiting to happen.

—Daniel E. Williams,
Sustainable Design: Ecology,
Architecture, and Planning

Water is Earth's most abundant resource—70 percent of Earth's surface is water—yet less than 1 percent is the readily accessible freshwater we human beings must have to survive. That's not much to meet the needs of the world's nearly 6.9 billion people, about 310 million of them in the United States.[1] Water was once without question a renewable resource, but that's not necessarily the case today. We've overused it, polluted it, drained it, and built up and over Earth's natural means to replenish it, especially when drought and changing climate are part of the equation. Even our approach to development has encroached on our ability to replenish our most needed, if not most valuable, resource. Parking lots, streets, and walkways pave out the water by interfering with the planet's natural ability to restock its freshwater supplies.

Where does Earth's water come from? Figure 1.1 illustrates the breakdown of Earth's water supplies.

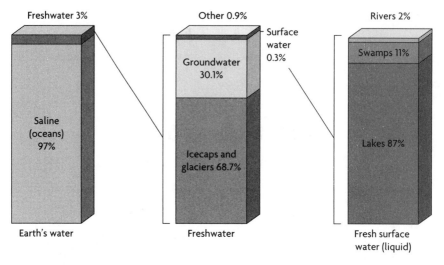

FIGURE 1.1 Distribution of Earth's Water
Source: U.S. Geological Survey.

According to a 2003 report from the U.S. General Accounting Office:[2]

- Oceans provide 97 percent—salty and undrinkable, and not usable for irrigation without costly desalination.
- Glaciers, polar ice caps, and groundwater contribute 2 percent—generally frozen and inaccessible.
- Surface water, including rivers and lakes, supplies 0.3 percent—generally accessible to satisfy freshwater needs that include energy, agriculture, and industrial and personal use.

The water crisis is global in scope, yet hits each of us personally. Not only do we need water to live, but we require massive quantities of it to produce energy, power industries, develop communities, and grow food. Even processing water for consumption requires huge amounts of water to produce the energy necessary for pumping, conveyance, treatment, and more. If water is in short supply, the costs of goods and services that rely on it go up, too.

For years, recycling bathtub and dishwasher water into the garden or onto the yard was the norm in water-short states like

California. Periodic restrictions on water use also are a way of life in other desert and mountain states like Colorado, Arizona, and New Mexico. But what about water shortages and restrictions in places like Pittsburgh, Pennsylvania; Boston, Massachusetts; Sarasota, Florida; Richmond, Virginia; Colstrip, Montana; Dodge City, Kansas; or Brentwood, New Hampshire?

OUR ENDLESS THIRST

The United States sucks up about 410 billion gallons of water every single day, 349 billion of which are freshwater. That's nearly 150 trillion gallons of water used every year in this country, according to the latest (2005) data from the U.S. Geological Survey (USGS), which compiles water-use numbers every five years.[3] To put water use in perspective, consider that the average American uses about one hundred gallons of water every day. That's about 36,000 gallons a year. (An average-sized backyard swimming pool might hold 18,000 to 20,000 gallons or more of water.) These numbers vary, depending on who's counting and where they're counting—geography, weather, personal use, and politics figure into the equation, too.

WATER FACTS

How much is one million gallons of water? According to the U.S. Geological Survey:

- A good-sized bathtub holds 50 gallons, so it takes 20,000 baths to equal one million gallons.
- A swimming pool that holds a million gallons would have to be about 267 feet long (almost as long as a football field), 50 feet wide, and 10 feet deep.
- For more, check out the USGS web site, http://ga.water.usgs.gov/edu/mgd.html.

Beyond those numbers, the amount of water our lifestyles require can be mind-boggling. It takes about five gallons of water just to produce a single gallon of gasoline! If that's not consumption enough, to pump, treat, and supply water to our homes also requires a tremendous amount of energy, which in turn requires lots of water. U.S. public water, sewer, and treatment facilities use about 56 billion kilowatt-hours of electricity every year, enough electricity to power more than 5 million homes, according to the U.S. Environmental Protection Agency's WaterSense program.*

*U.S. Environmental Protection Agency, "WaterSense," http://epa.gov/watersense/.

BIGGEST CONSUMERS

The biggest guzzler of water in the United States is thermoelectric power. The nation's power plants require 201 billion gallons of water daily to make steam to turn turbines, for cooling purposes, and more. That's up 3 percent from 2000. Irrigation is the second-largest water user, requiring 128 billion gallons every day. Domestic water use takes a distant third, with 44.2 billion gallons of water required a day to meet Americans' personal needs, followed by 31 billion gallons of water daily for industrial, mining, commercial, and aquaculture uses.[4] These numbers sound simple enough, but satisfying our nation's water demands is anything but simple.

IN OUR BACKYARDS

These numbers may sound mind-boggling and the idea of a water crisis here at home more like science fiction. After all, you turn on the tap and plentiful clean water comes out. Until recently the idea of a water crisis, if it existed at all, was someone else's problem. Few Americans other than environmentalists, a handful of government officials, farmers, and workers for world aid organizations like the United Nations actually paid much attention to the extent of the world's water issues.

Reality, however, is that the United States has tapped into, sucked up, and maxed out its once-abundant and replenishable supplies of freshwater on the surface and underground. In this country, too, thousands of people fall ill every year from waterborne diseases, according to data from the U.S. Environmental Protection Agency.[5]

Every place has its own issues with water, says Daniel E. Williams, FAIA (Fellow, American Institute of Architects) APA member

(American Planning Association), and author of *Sustainable Design: Ecology, Architecture, and Planning* (John Wiley & Sons, 2007). As an architect and urban and regional planner, Williams has studied water and planning issues in Hawaii, Washington, Florida, Colorado, Arizona, New York, Louisiana, and points in between.

For example, he says, "Colorado sells its water to Arizona, which is basically one desert selling it to the other. The aquifer that serves Tucson, Arizona, is down three hundred feet from water levels one hundred years ago. We're irrigating desert for our food source, because we've built all over the best arable land. A large percentage of our breadbasket right now is irrigated with million-year-old [ancient aquifer] water and water from Canada through the Columbia River and some of the other large transcontinental rivers. No city in the country supplies its own water within its geopolitical boundaries. So it's literally *stealing* water from another place, which isn't a problem as long as that other place doesn't need it. But once they need it, there's a big problem.

"It's a disaster waiting to happen," adds Williams, who currently is working on a book on urban design with climate change in mind.

"No one thinks the United States has a water problem," says Mike Hightower, water expert, environmental engineer, and Distinguished Member of the Technical Staff, Sandia National Laboratories. Based in Albuquerque, New Mexico, Sandia is part of the U.S. government's National Nuclear Security Administration, which studies water as a national security issue. "We have developed an extensive network of dams over the last hundred years, which has helped harness our vast fresh surface water resources. But we haven't built any major dams in the last twenty-five years, and we've maximized the use of our available surface water resources. To meet the growing demand for freshwater over the past several decades, we've gone to major utilization of groundwater in our aquifers [underground water supplies], and we've begun to overpump those."

A consequence of overpumping water from many of the aquifers that have been drawn down is that the water left over is often brackish or salty. Saltwater is heavier than freshwater and generally settles closer to the bottom of an aquifer, says Hightower. States that have had to deal with overpumping and salinity issues in parts of their aquifers include Arizona, North Dakota, Nebraska, California, Minnesota, Ohio, Illinois,

> ### WATER FACTS
>
> - **Source water:** Water from rivers, lakes, streams, or aquifers that has not been treated; is used to provide drinking water via wells or public water supply
> - **Groundwater:** Water found beneath the surface, such as in aquifers; can be the source water for various supplies
> - **Surface water:** Water found above ground, such as in lakes, rivers, and streams; can be a source for various water supplies

New York, Pennsylvania, Florida, Wyoming, Montana, South Carolina, Alabama, Louisiana, Arkansas, Texas, and Oklahoma, he adds.

Source Water

"We don't *run out* of water," says Eric Evenson, the USGS's National Water Census Coordinator. "Instead, the demand for it overstrips the supply. We still have a lot of water in various areas, but when we develop our population and our landscape, we don't always match up our water demands well with our water availability."

That's what's happened in Atlanta. The city's size and subsequent demand for water have mushroomed while what has been its primary water resource, Lake Lanier, hasn't. In fact, just the opposite has happened. In recent years, Lake Lanier, on the Chattahoochee River in north Georgia, has been under a veritable siege from drought, downstream demands from Alabama and Florida, and what Hightower calls "the giant sucking sound that is Atlanta trying to get all the water resources it can." Alabama and Florida want their water from the Chattahoochee, too, and the result is that the courts have a seat at the dispute table these days. (We delve more into water laws in Chapter 5.) The Chattahoochee forms a portion of the border between Georgia and Alabama, and then flows into the state of Florida, where it combines with the Flint River, subsequently becoming the Apalachicola River, which flows to the Gulf of Mexico. So there are three states and three groups—not including Atlanta— that want their share of what is a limited supply of water.

Water Shortages on Tap

Water woes across the country remain the norm, despite persistent on-again, off-again spotty wet and extreme weather. At least 36 states expect some kind of water shortage to continue through 2013 under normal climate conditions, according to a 2003 report from the U.S. General Accounting Office.[6] Factor in drought conditions that already affect areas of the United States, and that number climbs to 46 states!

The National Climatic Data Center reported that as of the end of August 2010, nearly 10 percent of the contiguous United States was affected by drought. The same report also said that nationally, summer 2010 was the fourth-warmest on record, with 10 Eastern states experiencing record warm summers— Rhode Island, New Jersey, Delaware, Maryland, Virginia, North Carolina, Tennessee, South Carolina, Georgia, and Alabama.[7]

Some other grim numbers to think about include:

- The temperature-related national energy demand for summer 2010 was the highest in 116 years. Keep in mind that energy is the single-largest daily consumer of water in the United States.[8]
- The Great Lakes is the largest freshwater lake system in the world and holds one-fifth of Earth's freshwater. Yet Lake Superior, the largest lake, hit record lows in August 2007.[9] The lake remained 13 inches below its average level as of August 2010, despite above-average precipitation for that month.[10]
- The United States had its fifth-driest December to February on record (2008–2009), with Texas recording its driest winter ever. Twelve states in the southern plains, Southeast, and Northeast had at least their tenth-driest January to February since 1895, according to the National Climatic Data Center.[11]
- Alaskans worry about water availability in some smaller cities because the rivers freeze above ground, making their water inaccessible for major portions of the year. Therefore, they have to store water during short spring and summer periods, which presents a number of water-quality issues, according to Sandia's Hightower.

- Along the East Coast, methyl tertiary butyl ether (MTBE), an air pollution–reducing chemical additive to gasoline whose use was banned in 2007, has been found in some groundwater.[12] In general, the MTBE ended up in groundwater supplies as a result of leaking fuel tanks or runoff contaminated by gasoline.
- In September 2010, Governor Steven L. Beshear of Kentucky sought federal disaster assistance for 35 counties hit hard by drought in the western and west-central portions of the commonwealth.[13]
- Tennessee also looked to federal relief in September 2010 for water-related natural disasters. In some parts of the state, the problem was too much water, as in flooding, and in others it was too little water that left parched fields and decimated crops. It's not the first year water has been an issue for the state. As of February 2009, 57 counties in the state had been designated natural disaster areas for agriculture as a result of persistent drought conditions.[14]
- In Massachusetts, minimal rainfall along the upper Charles River one recent year left Bostonians wondering just how much longer their water will last, especially if one factors in the pollution runoff problems in the Charles and Mystic rivers, which serve the area. Brockton residents, however, have an alternative. In fall 2008, the Aquaria desalination plant came online; it taps brackish water from the Taunton River.
- In Florida—the Southwest Florida Water Management District (including Sarasota) declared a water shortage emergency in spring 2008. Low water levels in the area's primary water supply source, the Peace River watershed, prompted that declaration. Indicative of the changing nature of weather and water (or lack of) accompanying it, Florida experienced its all-time wettest May in 2009.[15]
- Arnold Schwarzenegger, then governor of California, declared a state of emergency in February 2009, to help the state cope with three consecutive years of drought. According to estimates, the cost of drought at that time amounted to up to $644 million[16] and it's not over yet.

The Farm Connection

What if you don't live in any of those "drought" states, on Lake Superior, in Georgia or in Alaska, or any other area with water problems? Water issues in the Midwest, the Southeast, California, and beyond affect you, too. When the nation's breadbasket withers, so does your pocketbook. If Midwestern farmers and Florida, Oregon, and California producers and fruit growers don't have adequate moisture or access to water to grow their crops, that means less grain, less corn, less ethanol, fewer vegetables, smaller fruit crops, and higher prices for everyone.

Although today's technology enables farmers to grow more crops on less acreage, farming still takes water. The amount is "staggering," says Brad Rippey, an agricultural meteorologist in the Office of the Chief Economist, U.S. Department of Agriculture (USDA), and an author of the U.S. Drought Monitor, a weekly

WATER FACTS

Here are some U.S. irrigation facts from the U.S. Geological Survey:*

- About 61 million acres were irrigated in 2005.
- Withdrawals amounted to 128 billion gallons a day, down about 8 percent from 2000 and about equal to 1970 levels.
- The national average application rate was 2.35 acre-feet of water per acre. One acre-foot equals the amount of water needed to cover one acre with one foot of water.
- California, Idaho, Colorado, and Montana combined accounted for about half the total irrigation withdrawals.
- Massachusetts had the highest application rate in the United States—6.9 acre-feet of water per acre, "likely due to water-management practices in the many cranberry bogs," the USGS report stated.

*U.S. Geological Survey, "Estimated Use of Water in the United States," 23, http://pubs.usgs .gov/circ/1344/pdf/c1344.pdf.

Volume of water required to produce the following food:

- Wheat: 1,000–2,000 liters (260–520 gallons) to produce one kilogram of wheat.
- Other grain: 1,000–3,000 liters (260–780 gallons) of water to produce one kilogram of grain.
- Grain-fed beef: 13,000–15,000 liters (3,380–3,900 gallons) to produce one kilogram of grain-fed beef.

FIGURE 1.2 How Much Does It Take?

Source: Food and Agriculture Organization of the United Nations (FAO).

report that closely follows water availability across the country. (Check it out at http://drought.unl.edu/DM/monitor.html.)

Not all the water used by agriculture, however, disappears, adds Rippey. A percentage used for irrigation is returned to the ground. How much depends on the amount absorbed by the crops; the particular crop and its water needs; the type of soil (some soils, like clay, are not as porous as others are); the weather (more evaporation occurs in hot weather); and the climate—arid or not, for example. "As we find with everything to do with water, it's all more complicated than a simple question and answer," Rippey adds.

The Food and Agriculture Organization (FAO) of the United Nations estimates that the average individual requires two to four liters of drinking water a day (or slightly more than a half gallon to more than a gallon). Compare that with the 2,000 to 5,000 liters (520 to 1,300 gallons) of water it takes to produce that same person's daily food.[17] See Figure 1.2.

Though experts differ on the amount of water actually used to grow a particular crop or raise certain livestock, there's no question that it takes a lot. In fact, it takes so much that the amount of irrigated, cultivated cropland acreage in the United States has declined, in part due to dwindling groundwater supplies, according to a report from the National Water Management Center, part of the U.S. Department of Agriculture's Natural Resources Conservation Service.[18] The report points to the following areas of concern:

- The High Plains of Texas lost 1.435 million acres of irrigated, cultivated cropland over the period 1982–1997, according to the National Resources Inventory, which states, "Most of the loss is due to dwindling groundwater supplies from the Ogallala aquifer. Aquifer level declines have ranged from fifty feet to one hundred feet since 1980, with saturated thickness reductions of 50 percent."

- South-central Arizona has seen water table declines of 200 feet.
- In the southern section of the Central Valley of California (Kern, Kings, and Tulare counties), an overdraft of 800,000 acre-feet per year has resulted in declines of more than 200 feet in some areas.
- The Mississippi River Alluvial Aquifer in Arkansas has declined 100 feet in 90 years in the Grand Prairie region, and well yields have declined accordingly.

THE ENERGY PRODUCTION SQUEEZE

As we mentioned earlier, water also is essential to the production of energy, whether to cool drills as they search for oil and gas; to create steam to turn turbines that power generators and produce hydroelectric power; or to harness harmful carbon emissions in the air and inject them underground (carbon sequestration).

Adding to the stress that energy production imposes on water supplies, many options for alternative fuels are in geographic areas with limited and already stressed water resources. People tout oil shale, for example, as a great domestic option to meet future energy needs. Not only is exploiting oil shale a costly process that uses huge amounts of water to extract oil from rock, but the nation's big oil shale reserves are in arid states like Colorado, Wyoming, and Utah that are facing drought conditions and have no water to spare.

WATER FACTS

Water is required to produce fuel. Sandia National Laboratories, in Albuquerque, New Mexico, offers an estimate of how much water it takes to produce one gallon of various types of fuel:

- Conventional oil and gas: 1.5 gallons of water to extract and refine fuel
- Grain ethanol (biofuel): 4 gallons of water for processing
- Corn (ethanol): 980 gallons of water to irrigate corn
- Oil shale: 2–3 gallons of water to extract and refine product
- Hydrogen (synthetic fuel): 3–7 gallons of water for processing
- Coal to liquid synthetic fuel: 4.5–9 gallons of water for processing*

* Ron Pate, Mike Hightower, Chris Cameron, and Wayne Einfeld, "Overview of Energy-Water Interdependencies and the Emerging Energy Demands on Water Resources," SAND2007-1349C (Albuquerque, NM: Sandia National Laboratories, March 2007).

THE POCKETBOOK PINCH

Too little or too much water is a no-win situation from a pocketbook perspective. When Hurricane Ike threatened offshore oil-drilling platforms and onshore refineries in the Gulf of Mexico in 2008, oil prices shot upward. At the other end of the spectrum, drought the same year cost California farmers tens of millions of dollars in crop losses. (The April 2010 disastrous BP Deepwater Horizon drilling rig explosion was catastrophic environmentally to the Gulf of Mexico. But ironically, a report from the Obama administration points to no effect on oil prices long term.)[19]

The financial cost of water shortages is tough to measure. But, according to a 2003 GAO report,[20] the U.S. Department of Commerce's National Oceanic and Atmospheric Administration identified eight specific water shortages resulting from drought or heat waves over the past 20 years, each costing $1 billion or more. The largest shortage, costing $40 billion, hit the central and eastern United States in summer 1988.

The Federal Emergency Management Agency goes a step further and attributes an estimated $6 billion to $8 billion in losses annually to drought in the United States.[21] That's a lot of cash, no matter where you live (or how wealthy you are).

WASHED AWAY

Even cities and states that you would assume have no water issues grapple with major water troubles. In September 2010, ongoing drought mixed with declining surface and groundwater supplies led Pennsylvania to declare a drought warning in 24 of its counties, and a drought watch for another 43 counties.

The western Great Lakes, says Sandia's Hightower, including Minnesota and Illinois, and Ohio along Lake Erie have their own water issues. They face population growth and development, limited aquifer supplies, drainage patterns that preclude pulling water from the nearby Great Lakes (because what you take out generally has to be put back into the same drainage basin), and laws relating to water rights. In December 2008, the Ontario (Canada) Ministry of the Environment accused the city of Detroit of stealing water from the Windsor, Ontario, side of the border, a practice it claims has been going on since 1964.

Areas of the Great Lakes Basin—the lakes and the area surrounding them—including parts of Illinois, Indiana, Michigan, Minnesota, New York, Ohio, Pennsylvania, and Wisconsin, and the Canadian provinces of Ontario and Quebec—are facing water conflicts, agrees USGS's Evenson: "As you move farther away from the lake shores, where people are using aquifers/groundwater, you can find areas of conflicts, where either there is not a lot of [water] or moving water out of the basin raises concerns."

Across the country, antiquated infrastructure contributes to more water supplies washing away. That infrastructure includes crumbling water-delivery systems—think water main breaks—and outdated storm sewers and drainage systems that leach huge amounts of freshwater out of aquifers and carry it away as "waste" water.

Leaking pipes lose an estimated 7 billion gallons of clean drinking water every day, according to the "2009 Report Card for America's Infrastructure," from the American Society of Civil Engineers.[22]

Suburban sprawl, which creates impervious surfaces like pavement, parking lots, buildings, and roofs, adds to the runoff. Data from a 2002 study by the nonprofits American Rivers, Natural Resources Defense Council, and Smart Growth America,[23] paint a grim picture of billions of gallons of freshwater that used to soak into the ground and now runs off and away from major metropolitan areas in the United States. This lost infiltration ranges from a low of between 6.2 billion and 14.4 billion gallons a year in Dallas, to a high of between 56.9 billion and 132.8 billion gallons annually in Atlanta. The Atlanta runoff in 1997 alone was enough water to supply the average daily household needs of 1.5 million to 3.6 million people for a year! You can read more about this study in Chapter 3.

WATER WARS AND MORE

With all these concerns and problems, it's easy to understand how water wars can brew across North America, as well as around the globe. Such conflicts aren't new, but they become increasingly more contentious and tougher to swallow as demand intensifies and supplies grow scarcer in a country where the resource has been taken for granted for centuries.

WATER TALES

Is there a battle over water in your neighborhood, town, state, or region? You might be surprised at how pervasive disputes over this "liquid gold" really are. Here's an abbreviated version of just a few of the battles, large and small.

- *The Colorado River Basin.* Bickering has been the norm for decades over the water flowing (or not flowing) in this river, which is essential to the southwestern United States and northern Mexico. Among the states in the fray have been Arizona, California, Colorado, Nevada, New Mexico, Utah, and Wyoming. One of the most contentious fights occurred in the early part of the twentieth century over the proposed construction of the Boulder Dam, which is now called the Hoover Dam. The dispute pitted states against each other, private business interests against government ownership interests, the federal government against states' rights, and more. The dam was built, but the groups, including Mexico, still battle. Everyone wants a piece of the Colorado River's bounty, which isn't bountiful enough to go around.*

- *North Central Texas versus Oklahoma and vice versa.* The Tarrant County (Texas) Regional Water District and other Dallas-area water suppliers want water from Oklahoma and have gone to court over it. Oklahoma doesn't want to relinquish water across state lines and has passed legislation to that

*U.S. Department of the Interior, Bureau of Reclamation, Colorado River Compact 1922 (November 24, 1922), www.usbr.gov/lc/region/g1000/pdfiles/crcompct.pdf.

effect. The Texans claim the moratorium on interstate water sales is unconstitutional. The federal courts say it's not. But, said Oklahoma's Attorney General Drew Edmondson after the fending off the latest Texas attempts to gain access to Oklahoma's water, ". . . the water wars are not yet won."*

- ***Las Vegas and Snake Valley, Nevada, versus Utah and Snake Valley, Utah.*** The water-hungry city of Las Vegas wants water from the Snake Valley, a portion of which is in Utah, and is willing to pay big for it. The Southern Nevada Water Authority (Las Vegas) has plans to build a $3.5 billion, 285-mile pipeline to bring the water south. But it's not a foregone conclusion, as clamoring emanates from all sides on everything from ecological concerns to complaints that Nevada doesn't have the right to "Utah's water." Resolution of the dispute won't happen anytime soon in the courts or among the parties involved.†

- ***Virginia versus Maryland.*** This centuries-old fight over Potomac River water has its origins in colonial times. One modern battle found its way to the U.S. Supreme Court in 2003. The court ruled that Maryland owns the river bottom, but Virginia has rights to some of the water.‡

- ***Montana Groundwater Pollution.*** In Colstrip, Montana, in late spring of 2008, some residents ended up with cash in their pockets but otherwise high and dry after the courts ordered a power plant, PPL Montana, to pay millions of dollars for polluting the groundwater in nearby subdivisions.

*Oklahoma Water Resource Board, press release, "Federal Judge Rejects Bid for Oklahoma Water," July 16, 2010, www.owrb.ok.gov/news/newsPDF/Federal%20Court%20Rejects%20Tarrant%20Bid%20for%20Oklahoma%20Water.pdf; Oklahoma Office of the Attorney General, press release, "Federal Court Rejects Tarrant Bid for Oklahoma Water," May 3, 2010, www.oag.ok.gov/oagweb.nsf/3E67F1CEE13BC090862572B2005AD559/6C8419033E219970862577180079F549.

†Nevada Division of Water Resources, State Engineer's Home Page, "Interim Order No. 3 Notice of Postponement of Hearing and Scheduling," http://water.nv.gov/Hearings/waterhearing/snakevalley/documents/Int%20order%203.pdf; Nevada Department of Conservation and Natural Resources Division of Water Resources, "Informational Statement Regarding Southern Nevada Water Authority Water Right Application . . . ," http://water.nv.gov/hearings/waterhearing/SpringDryCaveDelamar/Pre-notice%20website%20press%20release.pdf.

‡*Virginia v. Maryland*, transcript of oral argument, U.S. Supreme Court, October 7, 2003, www.supremecourt.gov/opinions/03pdf/129orig.pdf.

THE NEW WAY OF LIFE

As hundreds of water battles rage on across the country, the reality of our water supply's future becomes clearer. The current and growing shortage isn't just an isolated case. Experts agree that even without factoring in drought or climate change, water shortages and disputes over this finite resource aren't likely to go away anytime soon. That's especially true in the West, where increasing populations exert pressure on already short water supplies. "There are so many demands on the limited resource that unless there is something like a widespread ocean water desaliniation effort, there simply is not going to be enough to go around," says USDA's Rippey.

More water restrictions have become a way of life across the country. Those restrictions take many forms, can be voluntary or mandatory, and can be as minimal as suggested use of low-water plumbing and cutbacks on outside yard watering or as stringent as laws that stipulate the type of plumbing required, forbid outdoor watering, limit indoor use, and impose financial penalties for violations and exceeding certain water use limitations.

The root of the water crisis historically is not so much population growth but change in the way we use water, change in per-capita water demand, says Joseph Dellapenna, internationally known water law expert and professor at Villanova University School of Law in Philadelphia. "Many people think, 'I don't take more showers than I used to; what do you mean by change in demand?' Everything we consume consumes water, and as our affluence has grown, we have all consumed a great deal more water.

"While the population of the United States doubled between 1950 and 1980, per capita water consumption increased sixfold during that same period. It has leveled off but at a much higher level," says Dellapenna. "Factor in continued population growth—at this very high level of water consumption—with global climate change or, as I like to call it, 'global climate disruption,' which wreaks havoc on our expected patterns of precipitation. Even with the most modest conception of global climate change, there are very good reasons to understand that dry areas like the southwestern United States are going to become larger and drier. And formerly humid areas are becoming drier. This is most obvious in the southeastern United States," he adds.

As a longtime resident of the Northeast, Dellapenna says this means that declared water emergencies are much more common today than 25 years ago in his part of the country, including New England and the mid-Atlantic states of Pennsylvania, New Jersey, New York, and Delaware. "In the Northeast we've generally been able to deal with relatively modest water restrictions like no watering of lawns or gardens, or no washing of cars outside of car washes that recycle their water. We haven't had the kind of crisis they've had in the Southeast, where there have been actual requirements of serious reductions in water usage. But we (in the Northeast) have had these problems on a recurring and frequent basis," Dellapenna adds.

TALLYING INDIVIDUAL WATER USE

For a moment, forget the massive amounts of water required by the nation, region, or state, and consider what water scarcity can mean to each of us at home. The amount of water used by a city's residents can be overwhelming.

The average person uses about 80 to 100 gallons of water every day at home—whether home is Portland, Oregon, or Portland, Maine—for flushing toilets, washing, bathing, and other household needs. That translates to each of us requiring about 36,000 gallons of water a year, according to estimates from the U.S. Geological Survey.[24] (Those usage numbers vary, though, depending on how they're determined, location, climate, and so on.) Atlanta, for example, with its population of 429,500 (not including its suburbs), would require nearly 15.5 billion gallons of water a year for home use, based on the 36,000-gallons-per-person figure. Let's look at more Atlanta numbers.

- The city's geographic area covers the equivalent of 87,000 acres.
- The USGS estimates that a one-inch rainstorm deposits 27,152 gallons of water on one acre, or 2.36 billion gallons of water on Atlanta.
- It would take more than six one-inch deluges to provide enough water to satisfy Atlantans' annual water needs just at home.

However, comparing per-capita water-usage numbers city to city, state to state, or coast to coast isn't always an accurate indicator of water use, experts agree. "It's sometimes erroneous, or at least much more complicated than the figures might indicate," says Ken Kramer of Austin, Texas, director of the 23,000-member Texas chapter of the Sierra Club.

"Generally, when people talk about per-capita water consumption, they mean the amount of water that goes through a particular water system, dividing it by the population of an area and coming up with a per-capita figure," says Kramer. "But different areas vary in terms of their water customers. For example, you may have one huge industry that sucks up a lot of water and actually buys it from a utility system. That can skew per-capita water consumption to a much higher figure than in, say, a bedroom community that doesn't use as much water.

"The best use of per-capita consumption is to note the differences in water use over time in one particular utility," adds Kramer. "How much a utility has reduced that figure over a particular period of time is a more reasonable comparison."

Pat Mulroy agrees. She's general manager of the Southern Nevada Water Authority, which serves a thirsty Las Vegas. "You can't compare city to city," says Mulroy. "It's all driven by climate and what you use your water for, and how the various entities calculate their per-capita consumption. The only usefulness of a per-capita number is benchmarking against yourself. Everything else gets you into a fruit cocktail argument that no one can penetrate. . . . It's a PR tool that's pulled out of the hip pocket every time someone wants to flog someone."

WATER FACTS

The amount of water that falls from the sky in an ordinary rainstorm might surprise you. Learn more about it with the U.S. Geological Survey's interactive calculator, found at http://ga.water.usgs.gov/edu/sc2.html.

Water Use Levels Out

The good news is that growth in U.S. water usage from the 1980s to today has remained relatively flat or slowed despite population growth. But, says USGS's Evenson, as the population continues to grow, water use will increase, too. Unfortunately, growth doesn't necessarily happen in places where water is abundant. "As our population grows, we tend to push into more and more areas where our water supplies are more and more marginal," Evenson says.

The number of gallons of water used per person per day actually declined between 1980 and 2000, according to Susan Hutson, lead author of the U.S. Geological Survey's report entitled "Water Use in the United States, 1950–2000" and contributing author to the 2005 report.[25] (See Figure 1.3.) That's in part due to the Clean Water Act, says Hutson. That statute limited water discharge for industries and thermoelectric power plants. Because of those limits, industries aggressively developed technology to recycle and reuse water, thus minimizing their water withdrawals. In 1950, it took an average 63 gallons of water to produce one kilowatt-hour of energy. In 2005, it took just about a third of that. Also contributing to the decline in water-use numbers is the loss of some of the nation's industrial base, which used to withdraw water directly from rivers and lakes, or drill private wells. New industries today often go to the public water supply instead, adds Hutson. Improvements in land irrigation technology mean less water is needed, and they make a difference in consumption numbers, too, she adds.

More highlights from that report:[26]

- Despite growing demands, the total amount of water withdrawn per day for all uses in the United States (410 billion gallons) is down from the peak withdrawal year of 1980.
- Fresh surface water accounted for 78 percent of total withdrawals in 1950, and only 66 percent in 2005. Saline surface water (mainly for thermo-electric power generation) makes up the rest.
- California, Texas, Idaho, and Florida accounted for more than one-fourth of all water withdrawals in 2005.
- California had the largest surface-water withdrawals in 2005, with irrigation as the primary use.

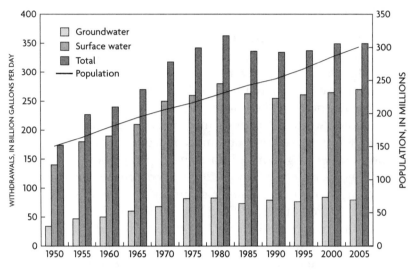

FIGURE 1.3 Trends in Population and Freshwater Withdrawals by Source, 1950–2005

Source: U.S. Geological Survey, "Estimated Use of Water in the United States in 2005," http://pubs.usgs.gov/circ/1344/pdf/c1344.pdf.

- Illinois, Texas, and Michigan counted on large quantities of surface-water withdrawals for thermoelectric power generation.
- Irrigation was the largest use of fresh groundwater in 25 states; California, Nebraska, Arkansas, and Texas were the biggest users.
- The bottom line: If it hasn't done so already, the battle for water will soon hit home for all of us.

WATER REALITIES

- The Earth is 70 percent water, but less than 1 percent is readily available fresh water humans need to survive.
- The United States withdraws 410 billion gallons of water a year.
- The average person uses about 80 to 100 gallons of water every day just at home. But those numbers can vary dramatically depending on how they're tallied, geographic location, and more.

- Adding to the stress on water supplies from energy production, many sources of alternative fuels are in geographic areas with limited and already stressed water resources.
- Water shortages and concerns have become a way of life from the Atlantic to the Pacific and from the Canadian border to the Mexican one, with various factions fighting over this shrinking twenty-first-century resource.
- Billions of gallons of wasted water wash away every year because of antiquated infrastructure.
- Water shortages hit the pocketbook hard. The National Oceanic and Atmospheric Administration identified eight specific water shortages resulting from drought or heat waves over the past 20 years, each costing $1 billion or more.

NOTES

1. U.S. Census Bureau, "U.S. and World Population Clocks—POPClocks," http://www.census.gov/main/www/popclock.html.
2. U.S. Geological Survey, "The Water Cycle: Water Science Basics," http://ga.water.usgs.gov/edu/watercyclesummary.html#global.
3. J. F. Kenny, N. L. Barber, S. S. Hutson, K. S. Linsey, J. K. Lovelace, and M. A. Maupin, "Estimated Use of Water in the United States in 2005: U.S. Geological Survey Circular 1344" (2009), http://pubs.usgs.gov/circ/1344/pdf/c1344.pdf.
4. Ibid.
5. Michael F. Craun, Gunther F. Craun, Rebecca L. Cauldron, and Michael J. Beach, "Waterborne Outbreaks Reported in the United States," *Journal of Water and Health* 4. Suppl. 06 (2006): 22–24, http://epa.gov/nheerl/articles/2006/waterborne_disease/waterborne_outbreaks.pdf.
6. U.S. General Accounting Office, "Freshwater Supply: States' Views of How Federal Agencies Could Help Them Meet the Challenges of Expected Shortages," Report to Congressional Requesters, GAO-03-514 (July 2003): 8, http://gao.gov/new.items/d03514.pdf.
7. National Oceanic and Atmospheric Administration, National Climatic Data Center, "State of the Climate National Overview August 2010, " http://lwf.ncdc.noaa.gov/sotc/?report=national.
8. Ibid.
9. National Oceanic and Atmospheric Administration, Great Lakes Environmental Research Laboratory, "Great Lakes Water Level Observations," http://www.glerl.noaa.gov/data/now/wlevels/levels.html.

10. U.S. Army Corps of Engineers Detroit District, "Monthly Bulletin of Lake Levels for the Great Lakes, September 2010," http://www.lre.usace.army.mil/_kd/Items/actions.cfm?action=Show&item_id=3887&destination=ShowItem.

11. "Climate of 2009: February in Historical Perspective," http://www.ncdc.noaa.gov/oa/climate/research/2009/feb/feb09.html.

12. News from the New Hampshire Department of Environmental Services, "MTBE Widespread in New Hampshire's Groundwater" (January 2, 2008), http://des.nh.gov/media/pr/documents/080102_mbte.pdf.

13. Kentucky.gov, press release, "Gov. Beshear Requests UDA Disaster Assistance," http://governor.ky.gov/pressrelease.htm?PostingGUID=7B6313482D-E371-47C7-A131-6244A57C72F2%7D.

14. TN.gov Newsroom, press releases, "Governor Bredesen Announces SBA Loan Program Available to Upper Cumberland Residents," http://news.tennesseeanytime.org/node/5913; "Bredesen Requests Presidential Disaster Declaration for 10 Counties," http://news.tennesseeanytime.org/node/5908; and "Bredesen Requests Federal Farm Assistance for Knox, Sumner Counties," http://news.tennesseeanytime.org/node/5893; "Bredesen Announces Federal Farm Assistance Granted for Giles and Macon Counties," http://news.tennesseeanytime.org/node/993.

15. NOAA Satellite and Information Service National Climatic Data Center/U.S. Department of Commerce, "U.S. National Overview: May 2009, National Climatic Data Center, Asheville, North Carolina (Updated 08 June 2009)," http://www.ncdc.noaa.gov/oa/climate/research/2009/may/national.html.

16. Governor of State of California, press release, "Gov. Schwarzenegger Joins Water March to Highlight Urgent Need to Improve California's Water Supply," April 17, 2009.

17. FAO Newsroom, press release, "Coping with water scarcity: Q&A with FAO Director-General Dr. Jacques Diouf," http://www.fao.org/newsroom/en/focus/2007/1000521/index.html; "FAO urges action to cope with increasing water scarcity," March 22, 2007, http://www.fao.org/newsroom/en/news/2007/1000520/index.html.

18. U.S. Department of Agriculture, "Long Range Planning for Drought Management—The Groundwater Component," http://wmc.ar.nrcs.usda.gov/technical/GW/Drought.html.

19. Rebecca M. Blank, Under Secretary for Economic Affairs, U.S. Department of Commerce, "Understanding the Impact of the Drilling Moratorium on the Gulf Coast Economy," testimony to U.S. Senate Committee on Small Business and Entrepreneurship, September 16, 2010, http://sbc.senate.gov/public/?a=Files.Serve&File_id=31647442-c186-48c8-b199-eea8fde7a0e1.

20. U.S. General Accounting Office, "Freshwater Supply: States' Views of How Federal Agencies Could Help Them Meet the Challenges of Expected Shortages," Report to Congressional Requesters, GAO-03-514 (July 2003), http://gao.gov/new.items/d03514.pdf.

21. Lifeng Luo, Justin Sheffield, and Eric F. Wood, "Towards a Global Drought Monitoring and Forecasting Capability," National Oceanic and Atmospheric

Administration's National Weather Service, 33rd NOAA Annual Climate Diagnostics and Prediction Workshop, Lincoln, Nebraska, October 20–24, 2008, http://www.nws.noaa.gov/ost/climate/STIP/33CDPW/Luo_33cdpw.htm.

22. American Society of Civil Engineers, press release, "2009 Report Card for America's Infrastructure," January 28, 2009, http://apps.asce.org/reportcard/2009/RC_2009_noembargo.pdf.

23. Betsy Otto, Katherine Ransel, and Jason Todd (American Rivers); Deron Lovaas and Hannah Stutzman (NRDC); John Bailey (Smart Growth America), "Paving Our Way to Water Shortages: How Sprawl Aggravates the Effects of Drought," *American Rivers, Natural Resources Defense Council, and Smart Growth America* (2002): 1–2, http://www.nrdc.org/ media/docs/020828.pdf.

24. U.S. Geological Survey, Water Science for Schools, "Water Q&A: Water use at home," http://ga.water.usgs.gov/edu/qahome.html.

25. Joan F. Kenny, Nancy L. Barber, Susan S. Hutson, Kristin S. Linsey, John K. Lovelace, and Molly A. Maupin "Estimated Use of Water in the United States in 2005," U.S. Geological Survey Circular 1344 (2009), http://pubs.usgs.gov/circ/1344/pdf/c1344.pdf.

26. Ibid.

CHAPTER

WHERE OUR WATER COMES FROM

A GLOBAL PERSPECTIVE

In so many places we simply take water for granted, and we think it will always be there.

—Steve Fleischli, former president,
Waterkeeper Alliance

The water crisis stretches well beyond U.S. borders. It's global, and its ramifications affect hundreds of millions of people.

The easiest way to describe the world water problem is that a billion people don't have access to safe drinking water, and 2.5 billion don't have access to adequate sanitation services, which leads to 2 million or so preventable deaths every year from water-related diseases. That's according to Peter Gleick, PhD, internationally recognized expert on global freshwater resources. He's also co-founder and president of the Pacific Institute for Studies in Development, Environment, and Security, an Oakland, California–based nonpartisan policy research group. "When someone tells me there's not really a water crisis and there's plenty of water, that brings me up short," says Gleick. "We have failed to meet basic human needs for many, and I describe that as a crisis."

Coping with the world's water scarcity is "*the* challenge of the twenty-first century," said Dr. Jacques Diouf, speaking at World Water Day celebrations in March 2007 in Rome. He's director general of the United Nations Food and Agriculture Organization (FAO). The National Oceanographic and Atmospheric Administration (NOAA), a U.S. government agency, describes the water issue as an "alarming decline in water supplies in certain regions of the United States and worldwide."

Let's look at some of the global statistics.

- Worldwide, 1.1 billion people don't have access to clean drinking water. That's 17 percent of the global population (World Health Organization, 2004).[1]
- Nearly 1.6 million people die annually because of unsafe water supplies; 1.8 million people die annually from diarrheal diseases, including cholera, with 88 percent of those deaths attributed to unsafe water supplies (World Health Organization, 2002).[2]
- As many as 135 million people could die from water-related diseases by 2020.[3]
- By 2050, 75 percent of the world's population could face a freshwater scarcity, according to UNESCO.[4]

Why should we in the United States, who live with access to *safe* water, care if parts of Africa or China don't have enough water, or if Europe faces drought, or if tens of thousands of people deal with polluted water every year? The answers aren't simple, and the issues aren't confined to the other side of the world.

THE WORLD IS THE STAGE

No one can live without water. Humans can live without food for up to a month or so, but can go without water for about a week at most.[5] That's the bottom line.

Imagine living in a place where you drill a well to provide your family with "safe" drinking water, only to discover that what comes out of the ground makes you sick. That's what has happened to hundreds of thousands of people in Bangladesh and

India, poisoned by arsenic-contaminated well water over several decades. In this case, the arsenic is naturally occurring. The situation repeatedly has been characterized as the worst such contamination in the world, according to the World Health Organization.[6] Moreover, India and Bangladesh are far from being the world's only trouble spots.

"People aren't dying of thirst per se," says Gleick. "They're dying of water-related diseases because they don't have safe water.

"It's not as though all the excess water in Canada or Norway or Iceland is of any assistance to the people in sub-Saharan Africa," adds Gleick. "It's too expensive to move water from one place to another."

Too bad. A town like Las Vegas, which has a typical rainfall of about 4 inches a year,[7] certainly could benefit from the 450 inches of rain that fall annually on Mount Waialeale, Hawaii.[8]

WATER FACTS

These figures come from the U.S. Geological Survey:*

- One gallon of water weighs 8.34 pounds.
- One cubic foot of water (7.48 gallons) weighs 62.4 pounds.
- One acre-foot of water (325,851.385 gallons) equals 43,450 cubic feet.
- One inch of rain equals 17.4 million gallons of water per square mile or 27,200 gallons (100 tons) per acre.

*USGS Water Science for Schools, "Water Properties," http://ga.water.usgs.gov/edu/waterproperties.html.

United Nations Initiatives

The United Nations has led the charge and set ambitious goals for clean water worldwide. In 1992, the United Nations designated March 22 as World Water Day, providing one way that nations, organizations, corporations, and individuals can highlight world water issues and discuss solutions to the problems. Initiatives and programs, small and large, have been developed as a result. The United Nations Millennium Declaration, passed by the General Assembly in 2000, affirmed the right of all human beings to safe water. In 2010, the United Nations General Assembly reaffirmed that right with a resolution that called on member states and international organizations to offer funding, technology, and resources to help poorer countries escalate efforts to provide clean, accessible, and affordable drinking water and sanitation for everyone.[9]

THE WORLD'S ECOSYSTEM

Whether you agree with the UN initiatives or not, and beyond the moral issue of access to water, the world is an ecosystem where climate, water, and weather are intertwined. What happens in someone else's backyard on the other side of the globe can and does affect what happens here.

Pacific to Atlantic and Beyond

Climate patterns, winds, and weather spread around the world and can affect U.S. water supplies. Drought in Asia, Africa, or the Middle East, for example, can generate dust storms, which end up as massive dust clouds in the atmosphere. Those clouds cross the Pacific Ocean on wind currents and rain down dust, pollution, and dirt across the United States and beyond. The United States even tracks those dust clouds.

Follow the current trail of dust around the globe online via NASA's Total Ozone Mapping Spectrometer, or TOMS (http://jwocky.gsfc .nasa.gov/aerosols/aerosols_v8.html).

Dusty trail. Those dust clouds also can affect rainfall amounts. They can clog the atmosphere and prevent minute raindrops in the atmosphere from growing large enough and heavy enough to fall to the ground as precipitation, impeding Earth's natural ability to replenish its water supplies. That natural process is known as the *hydrologic* or *water cycle,* which we'll detail later in this chapter.

In April 2001, a 1-million-ton dust cloud—one of the largest on record—developed in Asia, crossed the Pacific Ocean, left its trail across the United States from Alaska to Florida, and continued well out into the Atlantic Ocean before it dissipated. The extent to which that cloud interfered with precipitation isn't really known, but the cloud was very real. The dust storm was caused by winds from Siberia that kicked up dust in Mongolia and China! (Don't think all the dust comes from Asia. The United States exports its own dust and pollutants, which continue their eastward trek across the Atlantic Ocean.)[10]

Volcanic effect: snow in Alabama in July? Water quality expert and long-time hydrologist William R. Waldrop, PhD, remembers as a child hearing his grandfather talk about it. Says Waldrop, also president of Tennessee-based Quantum Engineering Corp., "My granddad talked about the snow in Northern Alabama, but everyone figured he didn't remember it right because he was just a kid at the time. I went back and checked, and the snow did occur a year after Krakatoa erupted on the other side of the world!" (The biggest volcanic eruption in modern history, Krakatoa exploded in 1883 in Indonesia, destroying most of the island and killing tens of thousands of people.)

Major volcanic eruptions *can* export weather effects across the globe as they spew gas and particulates high in the sky. Sometimes it's easy to pinpoint exactly what's happened, and sometimes it's not, says Brian Fuchs, a climatologist with the National Drought Mitigation Center at the University of Nebraska–Lincoln and an author of the U.S. Drought Monitor. The 2010 eruption of Iceland's Eyjafjallajökull sent plumes of ash into the air that clogged air travel in Europe for days. But, says Fuchs, its effect on world weather hasn't really been determined.

Volcanic eruptions can cool the atmosphere if the particles remain in the stratosphere, explains Douglas Le Comte, a meteorologist now retired from NOAA's Climate Prediction Center. "This rarely happens with eruptions outside of the tropics, so it is basically tropical eruptions that affect the climate, and this is over a several-year period. The last important climate-affecting eruption was Pinatubo in 1991 in the Philippines. This contributed to the cool U.S. summer of 1992," he adds.

When Pinatubo blew its stack in 1991, it sent nearly 20 million tons of sulfur dioxide into the stratosphere. Global temperatures dropped temporarily (1991–1993) by about 1°F, according to the U.S. Geological Survey.[11]

Climate trek. Just as the volcanic ash and the Asian dust cloud progressed around the globe, so do climate and weather patterns, which affect the amount of precipitation that falls, how it falls—as severe storms and flooding, snow, and so on—where it falls or doesn't fall, and its effect on water supplies.

What takes place in the Pacific Ocean, for example, affects the weather across North America, because weather moves from west to east, says Le Comte, also a contributing editor to *Weatherwise Magazine.* "Long-range forecasters pay attention especially to the tropical Pacific, where a tremendous amount of warm water and convection—warm air rising—means less stable air and more thunderstorms. That same energy rising into the atmosphere fuels the jet stream. All this and more contribute to storm and precipitation patterns across the United States."

That precipitation, of course, affects the amount of water in U.S. lakes, rivers and streams, recharge of aquifers, and so on. That's one reason El Niño and La Niña are perennial hot topics. Both concern changes in surface water temperatures in the Pacific and resulting weather patterns and precipitation across the United States. El Niño generally means a more active southern jet stream, which can lead to storms and more rain across the United States. La Niña, on the other hand, means colder ocean surface temperatures and drier U.S. weather.[12]

Climate change progresses around the globe, too. But the global connections between, for example, drought in Africa and the United States are inadequately understood, says Gleick. Nonetheless, he says, "Climate change is a very real problem that already has had impacts on water resources, reducing snowfall in mountainous areas of the United States like the Sierra Nevadas and the Rocky Mountains, and in the Himalayas on the other side of the globe. Climate change also affects the frequency and intensity of storms and drought, with increasing impacts in the southwestern United States and sub-Saharan Africa," Gleick adds.

How widespread is climate change? That depends on where and when you look. The central and southern regions of the United States had below-average temperatures in 2008, while the West, Southwest, and Northeast were above average. However, overall 2008 saw the coolest annual temperatures across the country since 1997, according to NOAA's National Climatic Data Center.[13]

The United States had some hot times in 2008, too, including its eighth-driest September on record, according to Le Comte, an expert in seasonal drought forecasting.

Globally, temperatures also are up. The year 2009 tied with 2006 as the fifth-warmest year on record for the Earth, according to a preliminary analysis by NOAA's National Climatic Data Center in Asheville, North Carolina. (Records date back to 1880.) More record or near-record numbers include the following:

- The average temperature for the decade (2000–2009), 57.9°F, was the warmest on record.
- July 2010 was the warmest on record with a worldwide land surface temperature, 1.85°F above the twentieth-century average of 57.8°F.
- January to July 2010, the year-to-date global combined land and ocean surface temperature, 58.1°F, was the warmest January–July period on record.
- In China, August 2010 was the warmest on record, according to the Chinese Meteorological Agency.[14]

Australia copes. The end of 2009 marked Australia's warmest decade on record, according to Australia's Bureau of Meteorology. Australians, however, seem to be learning to cope. It's the driest continent on the planet, says Richard Atwater, chief executive officer and executive director of California's Inland Empire Utilities Agency, a municipal water district in west San Bernardino County that serves the Chino Basin. Atwater visited Australia in June 2008 to see how Australians successfully handle their long-term drought. Brisbane, for example, is a gorgeous city with beautiful landscaping despite a bigger water shortage than Las Vegas or Southern California, says Atwater. But Australians landscape with much less water than in the United States. They don't allow the use of potable water to irrigate landscape, he says. Instead, they have 6,000-gallon tanks that capture rainwater running off the roof to irrigate plants and landscape.

"Like the Australians, we are all going to have to become more water-smart," Atwater adds. He and his utility already are national leaders in recycling and reusing wastewater, as well as producing renewable energy through methane gas and solar generation.[15]

The Polar Connection

One watery result of a warming planet is melting polar ice. As the north polar ice cap disintegrates, ocean levels and temperatures rise. That affects the location, amount, and frequency of precipitation. Arctic sea ice coverage in December 2010 was the lowest extent on record for the month since satellite records began in 1979. Arctic ice extent in December has declined 3.5 percent per decade (1979–2010), according to data analysis from the National Snow and Ice Data Center.[16]

Chilling Results

Whether climate change is natural, or is man-made and exacerbated by higher ozone levels and pollution, it is expected to decrease future water supplies. "Many water resource managers already are beginning to see it in the United States," says Sandia National Laboratories' Mike Hightower.

Climate change doesn't necessarily mean that total annual precipitation will be different, says water law expert Joseph Dellapenna. "But the pattern of precipitation will be different because more will fall as rain and less as snow. Even in the East, the real reservoir for summer water usage is the snowpack. If you get more rain and less snow, you're going to have less water in that reservoir. In states like California, it's quickly going to move into a crisis situation," he adds.

"In the East, the snowpack isn't as deep to begin with because the mountains are lower. So it may be more of a crisis than people realize, because they haven't focused on the snowpack aspect. But if you get more rain and less snow, the snow will melt sooner," Dellapenna adds.

Not everyone buys into the notion of "climate change." But change *is* happening. Spring already comes earlier than it did a century ago; at least from Mother Nature's point of view. "In New England and Canada, the maple sap from trees is running two weeks earlier than a century ago," says Dellapenna. "That's nature's way of adapting to climate change."

Earth's water cycle—evaporation and precipitation—has accelerated, too, increasing the amount of water feeding into the world's oceans, according to a new study by NASA and university researchers. The study, published in the Proceedings of the National Academy of Sciences, found that "18 percent more water fed into the world's ocean from rivers and melting polar ice sheets in 2006 than in 1994," with an average annual increase of 1.5 percent.

"That might not sound like much—1.5 percent a year—but after a few decades, it's huge," said Jay Famiglietti, University of California–Irvine Earth system science professor and investigator on the study, when the study was published in October 2010. "In general, more water is good. But . . . not everybody is getting more rainfall, and those who are may not need it. What we're seeing is exactly what the Intergovernmental Panel on Climate Change predicted—that precipitation is increasing in the tropics and the Arctic Circle with heavier, more punishing storms. Meanwhile, hundreds of millions of people live in semiarid regions, and those are drying up."[17]

Infrastructure woes. "Our dams, aqueducts, and water systems were built for yesterday's climate, not tomorrow's," says Gleick. "Tomorrow's climate will be different. We don't know the extent to which that infrastructure will be able to handle the changes that are coming."

Among potential infrastructure problems are reservoirs built the wrong size and in the wrong locations to capture enough water during runoff and wet times to serve the nation's needs in dry times, experts agree.

"Under prior climate conditions, our dam system stored adequate water for summer needs. But existing reservoir capacity isn't enough to capture the added runoff in what now is late winter, but by the end of the century, in terms of temperature, will be well into spring," Dellapenna says.

For example, says Gleick, "if runoff decreases in the Colorado River even 10 percent due to climate change—which is perfectly plausible—and we don't change the way we operate the river, the big reservoirs like Lake Mead and Lake Powell go dry very quickly." Adding to the stress on water supplies, "we have actually given away more water on the Colorado than it looks like nature will provide," says Gleick. "It's like a bank account where more [water, in this case] is going out than coming in. Pretty soon, your bank balance [or your reservoir level] dries up. You can't operate the system for the long term if you're spending more than is coming in."

Adding to the Colorado River's woes, a new study has found that dark-colored dust settling on the snow in the Upper Colorado River Basin robs the river of 5 percent of its water on average every year, and has been doing that to some extent since Europeans settled the West in the 1850s! The dust, caused by grazing, off-road vehicles, construction, and other soil disturbances, absorbs the sunlight, accelerates the snowmelt, and leads to earlier plant growth, and greater evaporation and transpiration.

The amount of water lost in the Colorado is nearly twice what the city of Las Vegas uses in a year, according to study co-author Brad Udall, director of the Western Water Assessment, a joint program of the Cooperative Institute for Research in Environmental Sciences, a collaboration of the University of Colorado and NOAA. "By cutting down on dust we could restore some of the lost flow, which is critical as the Southwestern climate warms," Udall said when the study was released in September 2010. It also was published in the *Proceedings of the National Academy of Sciences,* and funded by the National Science Foundation, the National Aeronautics and Space Administration and the Western Water Assessment.[18]

However, says Udall, cleaning up the dust won't solve Las Vegas's or the Colorado River's overuse and overallocation problems. "Say we're losing 5 percent of the water. In a perfect world, we could prevent 50 percent of that loss, which is more or less what Las Vegas uses in a year. But the overuse problems in the Colorado River basin won't be solved by restoring this relatively small amount of water."

There are numerous causes of the overuse, adds Udall, and the problems are shared by the Lower Basin states of Arizona, California, and Nevada. Population growth, climate change, outdated nineteenth-century water law, and cheap water in the American Southwest are huge long-term issues. "This is a typical, highly complex problem in the twenty-first century," says Udall. "It's not one problem. People want simplistic answers, yet these knotty problems defy simple solutions."

It's not just lower river water levels that are an issue with climate change. Hydrologist Waldrop points to the warming planet and rainfall and runoff patterns associated with it. Flood protection now is based on 100-year rainfall and runoff patterns. "If you change the rainfall patterns, those records are no longer accurate," he says. To some extent the problems of severe flooding in Pakistan in 2010 were exacerbated by the additional glacial melt coming down (and already maximizing) the rivers as a result of changing weather patterns."

> We should think of our aquifers as a savings account, not a checking account. You can live off the interest, but you can't live off the principal.
>
> —Daniel E. Williams, *Sustainable Design: Ecology, Architecture, and Planning*

Dried up. One frightening ramification of lower levels on the Colorado River is a "dead pool" Lake Mead. That's when the lake's water level falls below the intake for Hoover Dam. Power generation subsequently halts, and no additional water is released downstream. It's not science fiction. There's a 10 percent chance Lake Mead could be dry by 2014, and a 50 percent chance the reservoir levels will drop too low for hydroelectric power generation by 2017, according to a report from researchers at the Scripps Institution of Oceanography at the University of California–San Diego. The paper "When Will Lake Mead Go Dry?" by Tim Barnett, a research marine physicist, and David Pierce, a climate scientist, was released in February 2008. Among the culprits, according to the duo, are human demand, natural forces like evaporation, and human-induced climate change.

"We were stunned at the magnitude of the problem and how fast it was coming at us," said Barnett in releasing the report. "Make no mistake, this water problem is not a scientific abstraction, but rather one that will impact each and every one of us who live in the Southwest."[19]

Various scenarios consider the possibility that the lake will end up little more than what Pat Mulroy calls a "mud puddle" at different times in the future. Researchers offer different estimates—5 percent to 50 percent—on the chance of the dried-up scenario occurring in the next few years and beyond, says Mulroy, of the Southern Nevada Water Authority. "It's not a question of *if* it's going to happen; it's *when* it's going to happen. So what does that mean?" she asks.

"In the western United States, if you take the Colorado River out of commission, you are threatening all of Southern California, all the cities in Arizona, all of southern Nevada's economy, all of southern Nevada, and you're threatening some significant agricultural areas in California and New Mexico. So to have a better understanding of how climate change is going to manifest itself is of critical importance," says Mulroy. "Everything we've ever known, everything we've ever taken for granted or premises upon which we've calculated probability are meaningless. There is no rearview mirror."

Acting on climate change. People seem to agree that the climate is changing even if they don't agree on who is changing it or why, says Steve Fleischli, an attorney and former president of Waterkeeper Alliance, a New York–based international nonprofit activist group that concentrates on preserving and protecting the nation's waterways. "We need to understand the impact on water availability and supply. Then we have to look at how we use water and make sure we are managing it effectively and treating it as precious as it really is. Unfortunately, in some places we simply take it for granted and think that it will always be there," says Fleischli, who has been involved in water issues for more than a dozen years. He warns of inevitable problems if we continue to mismanage water, given increasing populations that take the resource for granted.

Water Scarcity and National Security

Beyond U.S. borders, water is emerging as a national security issue, too. "Water stress leads to conflicts, and conflicts lead to wars," says

Hightower, whose agency specializes in national security threats. "Globally, about 250 rivers cross international boundaries [including the Colorado and the Rio Grande along the United States' southern border, and the Columbia River in the north]. Water flowing from Turkey into Iraq and Iran, water flowing out of the Himalayas into Pakistan and India, and water flowing out of Tibet into China are all transboundary water resources."

The downstream water users feel the impact of upstream development because that development requires more water withdrawals. "Water resources, especially fresh surface water resources like rivers, are all potential transboundary conflict areas," adds Hightower, "especially when the natural resource is limited. The issue is not, 'Will there be conflicts?' There have already been hundreds of conflicts over water over the last two hundred years."

Conflicts are likely because as developing countries (and the United States) try to grow their economies, they require more access to freshwater, typically through their river systems. "If a country upstream is withholding water out of its river systems, that minimizes water that goes to the country downstream. Conflicts have and will continue to arise over that," says Hightower.

To put the potential for conflict in perspective, more than 3,800 unilateral, bilateral, or multilateral declarations or conventions concerning water exist today, according to data from UNESCO.[20] That includes 286 treaties, with 61 referring to the more than 200 international river basins.

U.S. border issues. Transboundary issues hit close to home. The Rio Grande, which runs along a big chunk of the United States' southern border with Mexico, has for decades spawned turmoil that has nothing to do with immigration and everything to do with water. There's simply not enough of it. In 2001, the Rio Grande actually stopped flowing into the Gulf of Mexico. Plain and simple, it ran out of water because of too many people and too many demands on its limited supply.

In recent years, cross-border skirmishes (of the legal and diplomatic kind) have only worsened as dry weather and drought have gripped much of the already water-short U.S. Southwest. In some cases, Mexico claims the United States isn't providing its full allocation of water as set forth in a 1944 treaty. In that treaty, Mexico agreed to provide the United States with a certain amount of water from the Rio Grande system and, in exchange, the United States agreed to give Mexico a certain number of acre-feet of water from the Colorado River. (One acre-foot equals the amount of water required to cover one acre with one foot of water.) More recently, Texas farmers have been disputing with Mexico and demanding millions of dollars in damages from that nation for withholding their water! In 2005, then–Secretary of State Condoleezza Rice even became involved, and supposedly ironed out the differences. But stay tuned. The fight isn't over. While the two countries have agreed on a water conservation plan for the upper Rio Conchos, none of the outstanding treaty issues has been resolved, says Stephen P. Mumme, PhD, a professor of political science at Colorado State University and recognized expert on water issues related to the Colorado River. The farmers have filed a protest as part of the North American Free Trade Agreement (NAFTA) dispute resolution process.

In another legal skirmish, Colorado River water is the prize. Mumme explains this United States–versus–Mexico fight: In Southern California's Imperial Valley, the fight emanates from the All-American Canal, an irrigation canal constructed decades ago to bring water from the Colorado to thirsty California. The United States unilaterally decided to concrete line the canal to eliminate seepage water—water that soaks into the ground and seeps under the international boundary with Mexico. Mexico has been using that seepage water for nearly 60 years, with as many as 30,000 poor farmers depending on it for their crops. The argument: The water in the canal belongs to the United States by treaty and can be better used to fulfill conservation requirements and deals among the Colorado River basin states, and with and within California itself.[21]

As one prominent water attorney said, "Basically this boils down to putting thirty thousand poor Mexican farmers out of work versus building maybe three thousand $2 million homes in the San Diego hills. Can that possibly be equitable?"

Of course not. Strong arguments can be made on both sides, but because the treaty allows the United States to take a particular position, Mexico may very well be out of luck, and many poor people may be hurt. That's the nature of water.

The United States faces long-running border disputes over water with its neighbor to the north, too. Canadian mining giant Teck Cominco has been battling the state of Washington and members of the Colville Confederated Tribes over pollution in the transboundary Columbia River. Though the case involves pollution, the real battle is whether U.S. environmental laws apply to a company outside the boundaries of the United States. In March 2009, a federal court ruled that Teck Cominco must pay the Colvilles more than $1 million in legal fees in the case. Legal wrangling aside, it's the water resource that suffers as a result.

For years, Teck Cominco's metal smelters discharged slag (fine-grained waste material) into the Columbia River from its plant in Trail, British Columbia, just upstream from the U.S. border. The practice stopped more than a decade ago, but years of slag remain upstream from the Grand Coulee Dam in Lake Roosevelt and are a source of what both Washington State and the Colville tribes say is pollution. Teck Cominco unsuccessfully tried to get the suit dismissed on the grounds that U.S. environmental law did not apply to it. The case went all the way to the U.S. Supreme Court, which declined to hear the appeal. The company has agreed to spend millions of dollars to study the effects of the slag. Meanwhile, the slag, the water problems, and the dispute remain.

Then there are the Great Lakes and their basin, which includes parts of eight states and two Canadian provinces (Illinois, Indiana, Michigan, Minnesota, New York, Ohio, Pennsylvania, and Wisconsin, and the Canadian provinces of Ontario and Quebec). The lakes' massive water supply seems to tempt the entire United States and some other thirsty parts of the world. Myriad schemes for taking water out of the lakes and towing, piping, draining, or shipping it to points around the world led, in part, to the final approval in 2008 of the Great Lakes–St. Lawrence River Basin Water Resources Compact (the Great Lakes Basin compact), touted as the way to protect, preserve, and conserve the lakes and their basin. That agreement provides a framework for managing and protecting Great Lakes water resources that includes a ban on most diversion of water from the Great Lakes Basin (the area that drains into the Great Lakes). Time will tell whether the compact works.

"There's a finite amount of water on planet Earth, and it's unequally distributed," says Mumme. He notes that in the United States—and to some extent Mexico—planning, property laws, and development have encouraged people to move to arid parts of the country. "So we're getting more claims on the water in this region just at the time when Mother Nature isn't cooperating. Whether that's due to systemic human-induced climate change or other processes independent of human beings is immaterial," says Mumme. "The fact is there is less water with much higher demand on it."

The 30th Parallel Conundrum

The U.S. Southwest, with its conflicts over the Colorado and Rio Grande rivers, is a typical example of how parties worldwide fight over a limited resource. Often, countries that face conflicts over water are mid-latitude nations located at about the 30th parallel above or below the equator. That includes the United States (especially the Southwest and Southeast), Spain, Morocco, southern parts of Europe, northern parts of Africa, the Middle East, Turkey, Iraq, Afghanistan, India, Pakistan, China, Australia, and northern Mexico.

"These areas already are or will be impacted generally by climate changes that create water shortages, because they are already short of water," says Hightower.

COUNTRIES MAKE UP SHORTFALLS

- Because the water table in the Mediterranean Basin has dropped, Barcelona, Spain, imports water from France to satiate its thirst, and not in bottled form, either.
- Turkey has made a business of exporting water. One of its big customers is Israel, where it's a water-for-arms deal.

Economic development, population growth, and climate change exacerbate the conflicts. "The other kicker in the equation is that not only do these nations need more water to develop, they need more energy—and that takes more water, too," Hightower adds.

Global Water Battles

The United States certainly hasn't cornered the market on concerns or disputes over water supply. Battles over water—whether clean drinking water, waste and wastewater, overuse or underuse, pollution, supplies, climate, or something else entirely—stretch across the globe. A few samples:

- Under a siege of sorts from climate change and pollution associated with humanity and industry, which threaten its future as a living lake, is Earth's largest freshwater lake, Lake Baikal. Located in Siberia near its border with Mongolia, the lake contains about 20 percent of the world's total fresh surface water, or as much as all the Great Lakes combined, according to data from the U.S. Geological Survey (USGS) (http://pubs.usgs.gov/fs/baikal/).
- In April 2008, Bangladeshis hit the streets to protest a water shortage. A falling water table and lack of electricity to operate pumps to access the deeper water brought protesters to the streets of Dhaka, the capital city.
- Chile and Bolivia have been in a verbal war over water in various rivers since the 1960s. At one point in 1978, the two even broke off formal diplomatic relations because of their ongoing dispute.

John Gandomcar has a unique perspective on the world water crisis. Born and raised in Iran, he's spent the last 25 years raising horses, ranching, and developing land in semiarid Colorado. "Water was abundant in Iran. The Euphrates River flowed down from the mountains above our town, Abadan. You could dig down five or six inches to reach plenty of water. The land was lush and green," says Gandomcar.

That's hardly the picture of some of Gandomcar's sagebrush-, dust-, and dry grass–blown acreage in Colorado. Fortunately, though, Gandomcar knows the value of water and water rights in the western United States, so his properties have access to the water he needs. "In Colorado, the rights control the water, and rights holders know when to use the water or waste it," says Gandomcar.

Not everyone, though, is as water wise. We talk more about water rights in Chapter 5.

WATER BASICS: A PRIMER

To understand why and how global water issues affect all of us, we first must understand how and where we get our water. Hydrology involves much more than calculating precipitation, evaporation, or how much water is underground or on the Earth's surface in our lakes and rivers.

The Water/Hydrologic Cycle

The movement of water, and the various forms it takes, is known as the *water cycle* or *hydrologic cycle*. It's a closed cycle—water doesn't escape into space. It has no beginning and no end; it's perpetual. (Figure 2.1 illustrates this constant state of motion.) Some bit of that water you just had in some form most likely came from distant places and distant climates at some time in the distant past. Earth, then, could be considered the ultimate recycler when it comes to water.

The amount of water on Earth doesn't change. It's the same today as it was yesterday, the day before, decades before, and millions of years ago. What does change, though, is the movement of that water; the form it takes (liquid, ice, or vapor); the degree of its purity, and its accessibility.

Let's look more closely.

Water supplies. The Earth's supply of freshwater basically comes from two sources: surface water and underground water.

- *Surface water.* Including rain, snow, ice, or variations of the three, this aboveground water fills our lakes, rivers, and streams.
- *Underground water.* Water found beneath the ground is known as an *aquifer*, and very often is pumped out of the ground for various uses. The top level of the aquifer is the *water table.*

An aquifer is like a constantly moving, giant storage basin—up to tens of thousands of square miles in area—of loose rocks, sand,

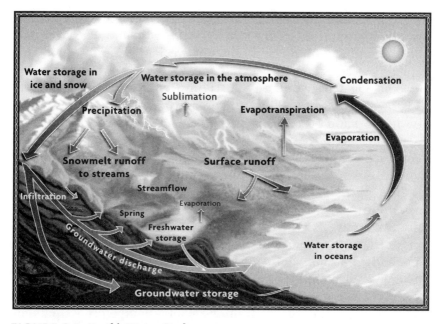

FIGURE 2.1 Earth's Water Cycle

Source: U.S. Geological Survey, modified from USGS illustration by John M. Evans, "The Water Cycle," at http://ga.water.usgs.gov/edu/watercyclesummary.html.

and gravel underneath us. It can be shallow, meaning it's relatively close to the surface and routinely replenishes itself with surface water runoff. That's a *renewable* aquifer. Other aquifers are deeper and don't refill or recharge as easily. Sometimes they may not replenish themselves in a lifetime or beyond. These are often called *prehistoric* aquifers. The last time many of these aquifers were fully recharged or filled, says Hightower, was in the Ice Age ten thousand years ago, when tons of ice melted all at once, saturating the ground down to the aquifers. "If we use that water up—and we are—it's not going to fully replenish itself until the next Ice Age," he adds.

The High Plains connection. One such prehistoric aquifer is the High Plains aquifer, which includes the Ogallala. The High Plains covers 174,000 square miles in parts of Colorado, Kansas, Nebraska, New Mexico, Oklahoma, South Dakota, Texas, and Wyoming and is the largest aquifer in the United States.[22] Data from the USGS and other organizations—public and private—agree that parts of the Ogallala and the High Plains are in trouble from overpumping for irrigation and other uses. More water is being taken out than put back in by natural replenishment (recharge), and supplies are depleting as a result.

"The Ogallala aquifer is not uniformly drawn down, either," adds USGS's Evenson. "There are areas where it's fairly dramatically drawn down and others where it's not."

The recharge process. Earth's water cycle can be complicated with many variations and interruptions. But basically, what goes up into the atmosphere eventually comes down to Earth, a portion soaks into the ground, and some rests on the surface, while the rest goes back up into the atmosphere, and the cycle begins again in perpetuity. See Figure 2.2.

"Recharge happens to aquifers from precipitation falling on the land and infiltrating into the groundwater system," says Evenson. "Where you have aquifers that are confined [deep aquifers with overlying clay layers like many along the Atlantic Coastal Plain], they get recharged from the leakage from the shallower aquifers above them.

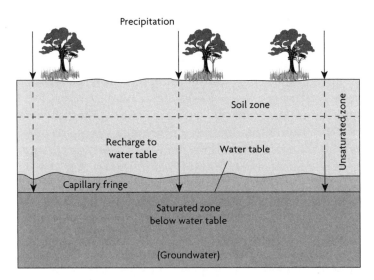

FIGURE 2.2 General Facts and Concepts about Groundwater

Source: Modified from U.S. Geological Survey illustration.

The leakage might be very small, though, with some parts of the country perhaps getting only a few inches a year of recharge."

Here's an excerpt of the USGS's "brief" (as in 20 pages of explanation) summary of the water cycle:

> . . . We begin in the oceans, since that is where most of Earth's water exists. The sun, which drives the water cycle, heats water in the oceans. Some of the water evaporates as vapor into the air. Ice and snow can sublimate (transform) directly into water vapor. Rising air currents take the vapor up into the atmosphere, along with water from evapotranspiration (a combination of evaporation and transpiration), which is water transpired from plants and evaporated from the soil. The vapor rises into the air where cooler temperatures cause it to condense into clouds. Air currents move clouds around the globe; cloud particles collide, grow, and fall out of the sky as precipitation. Some precipitation falls as snow and can accumulate as ice caps and glaciers, which can store frozen water for thousands of years.
>
> Snowpacks in warmer climates often thaw and melt when spring arrives, and the melted water flows overland as snowmelt. In some cases, especially in the arid West, dry winds blow over snow and ice and change it directly to the atmosphere as water vapor. Most precipitation falls back into the oceans or onto land, where, due to gravity, the precipitation flows over the ground as surface runoff. A portion of runoff enters rivers in valleys in the landscape, with stream flow moving water towards the oceans. Runoff, and groundwater seepage, accumulate and are stored as freshwater in lakes.
>
> Not all runoff flows into rivers, though. Much of it soaks into the ground as infiltration. Some water infiltrates deep into the ground and replenishes aquifers. Some infiltration stays close to the land surface and can seep back into surface-water bodies (and the ocean) as groundwater discharge, and some groundwater finds openings in the land surface and emerges as freshwater springs. Over time, though, all of this water keeps moving, some to re-enter the ocean, where the water cycle continues all over again.[23]

THE IMPORTANCE OF PLANTS IN THE WATER CYCLE

Plants play a vital role in Earth's water cycle because they *transpire* or, in effect, exhale water vapor into the air. To see this clearly, place a plastic bag over part of an ordinary houseplant or one outdoors, seal it, and leave it overnight. In the morning, you'll see the condensation (water droplets) on the inside of the bag, which are the result of the plant "exhaling" water vapor.

TODAY'S REALITY

This self-perpetuating water cycle has worked well for millions of years. But the cycle today is severely threatened globally and across North America. As climates change, populations grow and move, pollution increases, and needs and demands shift, the scarcity of Earth's essential resource—especially in a safe, clean state—has become a real threat. Let's look at reality in the United States.

All Dried Up

Reservoirs, rivers, and streams are drying up. We talked about what's happening along the once-mighty Rio Grande and the over-used Colorado. But beyond the major rivers of North America, the little streams and "washes" that once fed the land and the people have dried up, and many smaller rivers are being drained way down, too.

Historically one of the most important rivers in Massachusetts, the Ipswich, which meanders through the northeast portion of the state then out to sea, is a prime example of what can happen with overpumping. The river is only 45 miles long, but it's what Kerry Mackin calls the "lifeblood of the central part of Essex County not only because of its importance as a water supply but as the fluid that keeps everything alive here." Mackin is an environmentalist and executive director of the Ipswich River Watershed Association (http://ipswichriver.org). She's also been involved in water issues in the area for more than 15 years. "The river has been overallocated for public water supply, so for the past five decades large portions of the river have been pumped dry chronically."

The primary culprits have been lawn watering during dry summers, which coincides with naturally low flow, along with wells close to the river's edge that "literally pump the river dry," says Mackin. Adding to the problem, much of the water taken out of the river is removed from the watershed (drainage basin) completely. "Due to water rights granted in the early 1900s, about 20 million of the roughly 30 million gallons a day of water that have been allocated go to small cities south of the watershed, two of which are entirely and two partially outside the watershed . . . so the water never comes back. It's a huge net loss this watershed can't afford," says Mackin.

Shipped Out

Portions of Earth's largest freshwater system, the Great Lakes, feel the water squeeze, too. In fall 2007, the largest lake, Lake Superior, hit its lowest level since January 1926 and remains below long-term monthly averages (see Figure 2.3), according to data from the U.S. Army Corps of Engineers Detroit District.[24] The district regularly tracks water levels in all the Great Lakes. Beyond the issue of water for personal and industrial use, a dropping lake level has serious consequences for maritime shipping in Lake Superior.

Verbal battles between thirsty states lusting for Great Lakes water and those states and Canadian provinces with access to that water have brewed for years. Entrepreneurs looking to tap the water and state governments have entered the fray. In 1998, a Canadian firm, The Nova Group, won Ontario government approval to export close to 160 million gallons a year of Lake Superior water in tankers to water-short Asia! Public and private outcry on both sides of the border nixed that deal. The incident also led in part to the finalized 2008 Great Lakes Basin compact mentioned previously.

Farming Woes

Once abundantly productive land in the Midwest and West now barely produces enough crops to make farming worthwhile. Some fields today are dried and cracked at worst, and at best rely on "dryland farming" that depends on scarce rainwater that's getting scarcer.

FIGURE 2.3 Low Water Levels in Lake Superior: View of Duluth-Superior Harbor Looking Toward the High Bridge, 2007
Source: Jeff Gunderson/Minnesota Sea Grant.

The vast High Plains aquifer underlies portions of eight states, providing water for these major food-producing areas. But the ancient aquifer is being depleted. In fact, one of the great water wars of the future will be over the High Plains/Ogallala because people have overpumped it for years, says the U.S. Department of Agriculture's Brad Rippey. In some parts of the aquifer—areas of northwestern Texas, the Oklahoma panhandle, and southwestern Kansas—water levels are down more than 150 feet. Total water storage in 2005 in the aquifer was about 2,925 million acre-feet. (One acre-foot of water generally is enough for a family of four for a year.) That's down 253 million acre-feet, or 9 percent, since farmers began tapping the aquifer to meet their irrigation needs around 1950.[25]

Portions of the aquifer in west-central Kansas already are exhausted, adds Don Whittemore, senior scientific fellow at the

Kansas Geological Survey, University of Kansas–Lawrence. As a result, some areas have been hard-hit economically because farmers have had to turn to dry-land farming, which cuts crop yields. "The aquifer is like a bank account that's going down!" Whittemore adds.

Not only agriculture suffers from the drawdown. "Imagine standing at a well with a bucket, then dropping that bucket, filling it, and hoisting it up from one hundred feet compared with three hundred feet," says Whittemore. "You would be a lot more tired from the three-hundred-foot level because it takes more energy. It's the same with extracting water from the aquifer." Eventually cities will have a harder time pumping groundwater. As the aquifer is drained lower, it will become tougher, more expensive, and require more energy to pump out the remaining water.

WATER REALITIES

- The water crisis is global in scope. Worldwide, 1.1 billion people don't have access to clean drinking water, and about 1.6 million die as a result every year.
- Battles over quantity and quality of water brew around the world. Even the United States faces transboundary battles with its neighbors to the south and north.
- The world is an ecosystem. What happens in someone else's backyard on the other side of the globe, whether it's related to climate, water, or weather, can and does affect what happens here.
- Whatever causes climate change, it is a force that affects current and future water supplies.
- Even if climate change doesn't mean a change in total annual precipitation, changing weather patterns—earlier spring or less snow and more rain, for example—affect the ability of the nation's infrastructure to capture, store, and provide adequate water supplies.
- Earth's natural ability to replenish its water supply is called the hydrologic, or water, cycle.

- The amount of water on the Earth is constant. What changes is the form it takes—liquid, ice, or vapor; the degree of its purity; whether the pollution is natural or man-made; and the accessibility of that water.
- Fields in the Midwest and West that once produced abundant crops are parched and dry today as aquifers—underground water supplies—are drawn down and their supplies of water depleted.
- One aquifer in trouble in some areas is the vast High Plains, which includes the Ogallala aquifer.

NOTES

1. World Health Organization, "Water, Sanitation and Hygiene Links to Health," http://www.who.int/water_sanitations_health/publications/facts2004/en/index.html; World Health Organization, "Water, Sanitation and Hygiene Links to Health: Facts and Figures–updated November 2004," http://www.who.int/water_sanitation_health/factsfigures2005.pdf.
2. Ibid.
3. Peter H. Gleick, "Dirty Water: Estimated Deaths from Water-Related Diseases 2000–2020" (August 15, 2002), Pacific Institute for Studies in Development, Environment, and Security, http://pacinst.org/reports/water_related_deaths/water_related_deaths_report.pdf.
4. 2007 World Water Day, "Frequently Asked Questions," http://www.unwater.org/wwd07/faqs.html.
5. U.S. Environmental Protection Agency, "Drinking Water: Where Does My Water Come From," http://epa.gov/region07/kids/drnk_b.htm.
6. U.N. World Health Organization, Regional Office for South-East Asia, "Arsenic," http://searo.who.int/LinkFiles/Health_Topics_intro.pdf.
7. National Aeronautics and Space Administration Jet Propulsion Laboratory, California Institute of Technology, "GOES Weather Quiz: Explanations of the Answers," http://spaceplace.nasa.gov/en/educators/weather-quiz.pdf.
8. U.S. Geological Survey, "The Water Cycle," http://ga.water.usgs.gov/edu/watercyclesummary.html.
9. U.N. General Assembly (A/55/1000), letter dated June 25, 2001 from the Secretary-General to the President of the General Assembly, http://un.org/esa/ffd/a55-1000.pdf; UN News Centre, press release, "General Assembly Declares Access to Clean Water and Sanitation Is a Human Right," July 28, 2010, http://un.org/apps/news/story.asp?NewsID=35456&Cr=sanitation&Cr1.
10. Science@NASA, "All the World's a Stage . . . for Dust," June 26, 2001, http://science.nasa.gov/headlines/y2001/ast26jun_1.htm; "Dust Begets Dust," May

22, 2001, http://science.msfc.nasa.gov/headlines/y2001/ast22may_1.htm; National Oceanic and Atmospheric Administration, "Climate-Watch April 2001," May 23, 2001, http://www.ncdc.noaa.gov/oa/climate/extremes/2001/ april/extremes0401.html.

11. Chris Newhall, James W. Hendley II, and Peter H. Stauffer, "The Cataclysmic 1991 Eruption of Mount Pinatubo," USGS Fact Sheet 113-97, http://pubs .usgs.gov/fs/1997/fs113-97/.

12. National Oceanic and Atmospheric Administration, "Frequently Asked Questions about El Nino and La Nina," http://pmel.noaa.gov/tao/elnino/ faq.html.

13. National Oceanic and Atmospheric Administration, National Climatic Data Center, "State of the Climate National Overview August 2010," http://lwf .ncdc.noaa.gov/sotc/?report=national; NOAA's National Climatic Data Center, "Climate of 2008: December in Historical Perspective," http://www.ncdc.noaa .gov/oa/climate/research/2008/dec/dec08.html.

14. NOAA Satellite Information Center, "Climate of 2008 in Historical Perspective: Annual Report," January 14, 2009, National Climatic Research, http://www .ncdc.noaa.gov/oa/climate/research/2008/ann/ann08.html; NOAA, "NOAA: December Global Ocean Temperature Second Warmest on Record," January 21, 2010, http://www.noaanews.noaa.gov/stories2010/20100121_globalstats .html.

15. Inland Empire Utilities Agency, press release, November 17, 2008, http://www .ieua.org/news_reports/docs/press/PressReleaseTheCaliforniaSustainability AllianceAward.pdf.

16. National Snow and Ice Data Center, "Repeat of a Negative Arctic Oscillation Leads to Warm Arctic, Low Sea Ice Extent," (January 5, 2011), http://nsidc .org/arcticseaicenews/index.html.

17. "Proceedings of the National Academy of Sciences, October 4, 2010," http:// www.pnas.org/content/early/2010/09/28/1003292107.full.pdf+html; NASA, press release, "NASA Study Sees Earth's Water Cycle Pulse Quickening," October 4, 2010, http://www.nasa.gov/topics/earth/features/water20101004.html.

18. Cooperative Institute for Research in Environmental Sciences, press release, "Desert Dust Reduces River Flow, Says New Study," September 20, 2010, http://cires.colorado.edu/news/press/2010/dustonsnow.html.

19. Scripps Institution of Oceanography, press release, "Lake Mead Could Be Dry by 2021," February 21, 2008, http://scrippsnews.ucsd.edu/Releases/ ?releaseID=876.

20. U.N. Educational, Scientific and Cultural Organization, "Facts and Figures extracted from the 2nd United Nations World Water Development Report," http://www.unesco.org/water/wwap/wwdr/wwdr2/facts_figures/index .shtml.

21. Aquifornia, "Interagency consortium aims to raise All-American Canal awareness," http://aquafornia.com/archives/category/regional-water-issues/ all-american-canal.

22. V. L. McGuire, "Changes in Water Levels and Storage in the High Plains Aquifer, Predevelopment to 2005," U.S. Geological Survey, Fact Sheet 2007–3029, http://pubs.usgs.gov/fs/2007/3029/pdf/FS20073029.pdf.

23. U.S. Geological Survey, "The Water Cycle," http://ga.water.usgs.gov/edu/watercyclesummary.html.

24. U.S. Army Corps of Engineers Detroit District, Great Lakes 2008–2009 water level comparisons, http://www.lre.usace.army.mil/greatlakes/hh/datalinks/PrinterFriendly/DailyLevelsEnglish.pdf; "2008 Great Lakes Water Level Summary," http://www.lre.usace.army.mil/_kd/Items/actions.cfm?action=Show&item_id=5839&destination=ShowItem19.

25. U.S. Geological Survey Groundwater Resources Program, "Changes in Ground Water Resources and Storage in the High Plains Aquifer, Predevelopment to 2005 . . . ," http://pubs.usgs.gov/fs/2007/3029/pdf/FS20073029.pdf.

CHAPTER 3

THE DISAPPEARING ACT

*The fundamental way forward is to understand the problem of
water scarcity, and through a societal approach, deal with it.*
                    —Frank Richards, hydrometeorologist

What happened to America's water? It can't just vaporize—
or can it? As with everything else to do with water, the answer is
complicated, with many issues over time contributing to today's
squeeze.

The water shortage isn't consistent geographically, geologically, hydrologically, or historically. As we discussed in the last
chapter, the culprits include climate change and drought. Other
causes include geology and geography; changes in population
location and size, water, and land use (and overuse); strict water-
use regulations or lack thereof; infrastructure that is inefficient,
antiquated, and worn out; outdated water treatment plants that
can't handle twenty-first-century poisons; natural and man-made
pollution; and handling of water waste. Even the way we develop
land—paving vast areas and thereby keeping water from soaking
back into the ground to naturally replenish supplies—figures into
our water woes.

HISTORY REPEATS ITSELF

The series of dams developed in the United States over the last century enabled this country to establish a more extensive water supply system than many other countries. It also kept the United States about 50 years ahead when it came to maintaining adequate water supplies. Other parts of the world with water problems similar to those of the United States simply started earlier, says Sandia National Laboratories' Mike Hightower.

The world water shortage started in the 1940s and 1950s, especially in the Middle East, which didn't have river and dam systems to utilize and store water. Those countries followed a natural progression when it came to tapping what water resources they had. First, says Hightower, they looked to surface water to meet their needs, then groundwater, and, as they used that up, they turned to widespread water conservation, major water reuse projects, and coastal desalination. "We're seeing the same progression in the United States. We built lots of dams in the 1930s and '40s to develop our fresh surface water supplies, so we have more resources than other parts of the world. But in the 1970s, we turned to greater groundwater pumping. Now we're using up those resources and have limitations on our fresh surface water supplies. We're going to have to make extensive changes in the next 25 years to address those emerging water shortages," Hightower adds.

Cities, towns, municipalities, and states across the country—even the federal government—have their own issues and ways of dealing with water. The themes, however, are similar: too much demand for too little of the resource, quarrels over ownership of or rights to the water, how the resource is being used (or not), and pollution related to it. That leads to the world water equation:

Growing demand + Limited supply = Shortage and conflict
over what's left

Let's look closer at some of the causes and effects.

CLIMATE CHANGE

Obviously, climate—wet or dry and degrees in between—has a huge impact on the availability of water. Arid climates have less water, less rainfall, less availability of surface water, and less replenishment of groundwater than do humid ones.

The Real Issue?

People often confuse weather and climate, and what they're experiencing day to day with what is happening worldwide. Weather is a short-term occurrence and climate long-term, says Tennessee hydrologist William Waldrop. "We have a cold snap, for example, and therefore 'there's no such thing as global warming.'" Of course, the warming of the planet is well-documented, adds Waldrop. In the 1980s when he was at the Tennessee Valley Authority, Waldrop worked on studies that examined the potential effects of extreme climate on water resources. "We couldn't call it 'climate change' back then (because it wasn't politically correct at the time)."

Climate change is not new. It has always occurred, just like Earth's recycling of its water supply. Factor global warming in with it and things begin to really change (Figure 3.1). "It is society's lack of ability to adjust to these changes that creates problems," says Frank Richards, hydrometeorologist, now retired from the National Weather Service's Hydrologic Information Center in Silver Spring, Maryland. Richards specializes in precipitation and how it behaves on the ground.

FIGURE 3.1 Grinnell Glacier, Glacier National Park, 1910 (left) and 1997 (right)
Source: Glacier National Park Archives, F. E. Matthes and Lisa McKeon.

When change is slow enough, societies generally can adapt, says Richards. The Gobi Desert, for example, is in one of the 30th-parallel geographic regions mentioned in Chapter 2. Its inhabitants include wandering nomad tribes that have learned to cope with life in an arid environment where water is scarce. They move around with the shifting sands—and water.

Unfortunately, back in the United States—and a good chunk of the rest of the world—it's not quite as easy to adapt to less water. Much of the U.S. West is arid or semiarid. Mix that with the area's booming population growth, and water supply becomes a serious issue. Factor in past, present, and future drought, and the situation hits crisis proportions. Nightly television news reports often offer rainfall totals alongside daily temperature highs and lows.

"While climate change will have a significant impact on water resources, society's ability to effectively address water issues will depend on science—both to understand the problem, and to develop ways to better manage the Earth's valuable water resources," Richards says.

The mega-issue in the United States is that since the Jimmy Carter and Ronald Reagan presidential years, with a few minor exceptions, the country hasn't built any (major) water storage systems, according to A. Dan Tarlock, water issues expert, author, consultant to government agencies, and professor at Chicago–Kent College of Law in Chicago, Illinois. "We're not building any dams or reservoirs anymore for environmental or economic reasons. There has been talk that because of climate change, you have to build more storage. That would be the mega-development that could change things," adds Tarlock.

But, says Richards, "Climate change, when looked at broadly, may not be quite as daunting when you focus on its positive aspect. As with most change, there are winners and losers. If we have less water in one part of the country, precipitation may actually increase in another part of the country."

As an example, Richards points to 1993, when the Midwest experienced a significant flood at the same time the Southeast was suffering a severe drought. The average rainfall across the country, however, was just about normal. Therefore, Richards says, the challenge is accommodating the variability in the water cycle's timing and location and, more importantly, its variation from what we see and are used to as "typical."

Understanding Drought

Scientists and weather experts use many criteria to define drought, including the amount of moisture in the soil; how much precipitation has fallen; the volume—or lack thereof—of water flowing in streams; the content of reservoirs, that is, how full or empty they are; and how current conditions compare with past ones. The level of drought severity varies based on numbers and scientific calculations. Basically, however, drought is a moisture deficit over a broad area and usually over a prolonged time. That deficit results in shortfalls in the water needed to meet the needs of people, business, and industry.

Climate prognosticators currently see portions of the Southwest, Southeast, Northeast, and Hawaii experiencing varying degrees of drought. As of the end of November 2010, about 12 percent of the contiguous United States was characterized as in moderate to extreme drought, according to the official U.S. Drought Monitor (http://www.ncdc.noaa.gov/sotc/drought/2010/11). The United Nations labels much of the southwestern United States as being in a state of "physical water scarcity." They define this as a condition where "water resources development is approaching or has exceeded sustainable limits."[1]

Even though some places experience a month here or a month there with plentiful rainfalls, it's the short-term Band-Aid versus the long-term solution. Drought-plagued Huntsville, Alabama, for example, received 10 inches of rain in December 2008. It was the first time the area had that much precipitation in one month since 2003, and it was enough to pull it out of a more-than-two-year drought. In fact, December that year provided so much rain that northern Alabama and southern Tennessee had to deal with flooding concerns in January 2009. In April 2009, Tampa, Florida's primary reservoir was all but dried up. The next month Florida recorded its all-time wettest May.

More good news resounded on the drought front in the last quarter of 2008. U.S. Department of Agriculture (USDA) meteorologist Brad Rippey pointed out that as of October 2008, the United States was at its lowest coverage of drought since October 2005, thanks to the large number of tropical storms that hit the mainland United States in 2008. "I would argue," adds Rippey, "that 1998 to 2005 was the third most significant drought we've seen since the beginning of the twentieth century [behind the Dust Bowl of the 1930s and the mid-1950s]. We began to pull out of the really severe drought about 2005, depending on the location."

Recovery is relative, though. It may have come too late for many farmers, communities, and dependent businesses, or not at all in the nation's heartland. That's not overdramatizing, either. It's reality.

The end result of climate change is that its macro-level influence on agriculture—the big picture—could well be muted, says Richards. "But locally, the impact on individual farmers may be catastrophic in the drier areas, and it may be beneficial in those that start to get wetter or warmer as a result of changing climate."

Global Warming

Americans may simply have to get used to warming temperatures and learn to live with them.

"Climate models show that the West is one area where global warming has had an impact on temperatures—not everywhere in the West, but [in] much of it—and that's expected to continue," says meteorologist Douglas Le Comte, a former National Oceanic and Atmospheric Administration (NOAA) drought specialist, an expert on seasonal drought forecasting, and contributing editor to *Weatherwise Magazine*. "The big issue in the future is if there will be enough rain and snow in those areas to offset the impact of the higher temperatures," he adds.

Other areas across the United States aren't immune from drying climate conditions, either. More than 1,100 counties—one-third of all counties in the United States—face higher risks of water shortages by mid-century as the result of global warming, according to a July 2010 report from California-based consultants, Tetra Tech, for the Natural Resources Defense Council. "More than 400 of these counties will face extremely high risks of water shortages," the report warns.[2]

Can the doom and gloom be that bad? Perhaps it's simply a matter of adjusting upward temperature expectations. Consider how the National Climatic Data Center characterized the 2007 drought in the context of history:

> The most extensive national drought coverage during the past 110 years (the period of widespread reliable instrumental records) occurred in July 1934 when 80 percent of the contiguous United States was in moderate to extreme drought. Although the current drought and others of the twentieth century have been widespread and of lengthy duration, tree ring records indicate that the severity of these droughts was likely surpassed by other droughts, including that of 1579 and the 1580s over much of the western U.S. and northern Mexico.[3]

Check out the USGS WaterWatch site for the current status of water resources in your area and current drought conditions (http://water.usgs.gov/waterwatch/).

Strained Water Resources

This kind of climate change strains U.S. water resources. That's the official assessment of the situation by the U.S. Environmental Protection Agency's (EPA's) Intergovernmental Panel on Climate Change. In 2007, that panel used computer models of Earth's climate system to explore the effects of climate changes. The bottom line of all this research was that there is less water of lesser quality for more people to share. The panel's findings suggest a few of the possible ramifications of climate change on North America.[4]

In general:

- Warmer weather will affect seasonal availability of water by increasing evaporation and reducing snowpacks. The Columbia River and other heavily used water systems in western North America will be especially vulnerable.
- Higher variability of precipitation will make water management more difficult.

By region:

- *Alaska:* Glacial retreat/disappearance in south, advance in north; impacts on flows, stream ecology
- *Pacific Northwest:* Earlier snowmelt, more frequent rain or snow, changes in seasonal stream flow, possible reductions in summer stream flow, and reduced summer soil moisture
- *West and Southwest:* A likely reduction in snowpacks and seasonal shifts in runoff patterns, possible declines in groundwater recharge, and reduced water supplies
- *Midwest:* Big declines in summer stream flow and increased likelihood of severe droughts
- *Great Lakes:* Possible lake-level declines, reduced hydropower production, shallower ship channels, and reduced water quality
- *Northeast:* Generally adequate water supplies but limited total reserve capacity, decreased snow cover amount and duration, possible large reduction in stream flow, and possible elimination of bog ecosystems
- *Southeast, Gulf, and Mid-Atlantic:* Possible longer droughts, increases or decreases in runoff/river discharge, and increased flow variability

The very real and very local effects of climate change were further detailed in a more recent study from the U.S. Global Change Research Program, a consortium of 13 government science agencies and several major universities and research institutes. The report, "Global Climate Change Impacts in the United States," was released in June 2009. *"Observations show that warming of the climate is unequivocal,"* the study says in its summary. "Climate-related changes have already been observed globally and in the United States. These include increases in air and water temperatures, reduced frost days, increased frequency and intensity of heavy downpours, a rise in sea level, and reduced snow cover, glaciers, permafrost, and sea ice. . . . These changes will affect human health, water supply, agriculture, coastal areas, and many other aspects of society and the natural environment."[5]

Weather Woes, Too

It's easy to understand how drought directly restricts supplies of water. Believe it or not, severe weather—including flooding, thunderstorms, freezing rain, and snow—can cause problems, too.

Ask the residents of Ames, Iowa, about the hazards of too much water. They're experts after August 2010 flooding left the city out of water—drinking water, that is. For several days, the city's 55,000 residents, including the University of Iowa, were under a boil-water advisory or had to rely on bottled water.

In Missouri, a state typically flush with water, residents ran out of drinking water after an ice storm knocked out electricity to a widespread area. The state's department of natural resources had to warn residents not to drink the water because the pumps that moved water through the state's water systems to keep out pollutants and contaminants were powered by electricity. Without operational pumps, and with extensive broken pipes due to the cold, drinking water supplies became suspect. It's a catch-22 even when there's enough water.

Societal Issues, Too

The fundamental issue, then, is how does the United States adjust to climate change? Hydrometeorologist Richards, along with many others, think the solution lies in more effective management of water, conservation, rethinking how and where we grow specific crops, and more. The issues are challenging and complex. Scientists need to provide information and tools to ensure all stakeholders are able to make informed decisions. "Environmental factors may be less tractable than the societal issues," Richards says.

"Society doesn't have a lot of ability to substantially modify climate conditions in the short term. And society can have difficulty staying with a program with long-term benefits and short-term pain, so it's important they understand the urgency."

Besides, towing icebergs from far-off Alaska to a parched Southern California or transporting millions of gallons of water from a flooded Texas to a dry Kansas doesn't make financial sense today. "Water will have to get much scarcer for that to happen, because the infrastructure costs are daunting," says Richards. "And frankly, I don't think water costs near enough to support that as an economic model."

"We put more and more people where water isn't and then complain we have water problems. Is that the climate or is that society?" Richards asks. "Reality is that in today's society we now require water to be allocated for environmental purposes, and we place more demands on water in terms of how we use it. So, while we've changed the uses of water, we haven't necessarily changed the amount of water we have."

Las Vegas, Nevada, is a high-population center in an arid climate with tight water supplies. However, the city's water utility, the Southern Nevada Water Authority, also is on the leading edge of learning to cope. With the level of its major water source, the Colorado River–fed Lake Mead, down well more than one hundred feet since January 2000 and population up by more than four hundred thousand since 2002, you would assume water usage has sharply increased. Actually, it's just the opposite. The Southern Nevada Water Authority reports that southern Nevadans' annual water consumption decreased by nearly 26 billion gallons between 2002 and 2009.[6]

Antiquated Infrastructure

Changes in water usage and climate change aren't the only things sucking up water supplies. Worn-out and inefficient pipes and waste-water systems in many of the nation's cities add to the drain and strain. Every single day, millions of gallons of good water drain away along with the bad. In fact, as talked about in Chapter 1, leaky pipes account for the loss of billions of gallons of clean drinking water every day, according to the American Society of Civil

Engineers' "2009 Report Card for America's Infrastructure." The United States' drinking water systems earned a D– grade from the industry organization.[7]

Good Down the Drain with the Bad

More than half of the "wastewater" that runs off and away via wastewater infrastructure in many cities actually is potable water that drains or leaks into the systems and then is carried away forever. In the Boston metropolitan area, for example, a recent study revealed that in a typical year with normal rainfall (45 inches), more than 60 percent of the water treated at the Massachusetts Water Resources Authority's Deer Island wastewater facility was actually potable water or rainwater that ended up in the system.[8] These statistics were reported by Robert Zimmerman Jr., who is executive director of the nonprofit Charles River Watershed Association, an advocacy group that serves the greater Boston area. Deer Island, in the middle of Boston Harbor, collects wastewater from forty-three communities, and then disposes of it nine and a half miles out in Massachusetts Bay at a rate of 380 million gallons a day. Of that volume, 230 million gallons a day of *good* water heads out to sea, rather than back into the aquifers. "That's about 85 billion gallons of water a year, which is the equivalent of the entire flow of the Charles River," says Zimmerman, who advocates decentralized wastewater treatment facilities as an environmentally sound way to make the most of freshwater supplies. Boston's water pipes leak, too. Some of the pipes still in use in Boston date back to the late nineteenth and early twentieth centuries and are made of brick and mortar. The mortar has cracked and has "enormous" leaks, Zimmerman adds, because water, after all, is attracted to where the pressure is the lowest. To get a better idea of how groundwater finds fissures and weak links in the pipes, Zimmerman offers the following scenario: Picture a bathtub with a pipe that comes out of the wall, goes through the tub, and then exits out again onto the floor of the room. If you fill the tub up with water, then punch a hole in the pipe, where does the water in the tub go? It seeks the point of least resistance, blows into the hole in the pipe and out onto the floor of the room.

Replacement: The Only Option

A big part of the solution would seem to be a no-brainer: replace the pipes. The only problem is the cost to do that is gargantuan. When many of the pipes were laid underground years ago, the ground above likely had few structures or infrastructure. Today it's a very different landscape, creating huge complications for any replacement projects.

Not only Boston is losing its drinking water to leaky pipes. Many of the invisible underground pipes that crisscross America's metropolitan areas are wearing out or are already worn out. That's a big reason contributing to the growing numbers of broken water mains across the country. These pipes have lived a long and useful life, but now they need to be replaced. Cost estimates for that replacement range from $485 billion to nearly $1.2 trillion for both drinking water and wastewater infrastructure over the next 20 years, according to estimates from the EPA, the Congressional Budget Office, and the Water Infrastructure Network, a consortium of industry, municipal, state, and nonprofit associations.[9]

> The American drinking water infrastructure network spans more than seven hundred thousand miles—more than four times the length of the total National Highway System, according to EPA estimates.*
>
> *"The True Cost of Water," *The WaterSense Current*, no. 8 (Fall 2008), http://epa.gov/watersense/about_us/fall2008.html.

Another industry report from the 60,000-member nonprofit American Water Works Association (AWWA) pegs the cost of replacing only drinking water infrastructure at about $250 billion over the next 30 years. That report, by AWWA's Water Industry Technical Action Fund, involved an analysis of 20 major utilities.[10]

Here are some of the more sobering findings of the report:

- The oldest cast-iron pipes, which date to the late 1800s, have an average life expectancy of about 120 years. Pipes laid in the 1920s last an average of 100 years, and pipes laid after World War II last about 75 years.

- The cost to consumers is huge. The government will have to be involved in bearing the expense. Nonetheless, some of the burden will fall on consumers. On average, the replacement cost of water mains is about $6,300 per household. Factoring in the cost of water treatment plants, pumps, and more, the replacement cost values rise to just less than $10,000 per household.
- Demographic shifts affect economics of reinvestment. In some older cities where population has declined, the per-capita replacement value of water mains is more than three times higher than the average in the sample.
- Costs are constantly increasing. By 2030, the average utility will have to spend three-and-a-half times as much on pipe replacement due to wear-out as it spends today, and three times as much on repairs as pipes age and become more prone to breakage.
- In the 20 utilities studied, infrastructure repair and replacement will require additional revenue totaling about $6 billion above current spending over the next 30 years. This ranges from about $550 per household to almost $2,300 per household over the period, not including the cost of compliance with new regulations.
- Household impacts will be two to three times greater in smaller water systems ($1,100 to $6,900 per household over 30 years) because of the disadvantage of small scale and the tendency for replacement needs to be less spread out over time.

Get the Lead Out

Another part of the problem is that for many years the standard pipe used to transport drinking water throughout the United States was made of lead. Of course, those pipes were in place before studies revealed the dangers associated with lead (including impaired intelligence and physical development in children, and increased risk of kidney problems, high blood pressure, and brain damage in adults). That revelation engendered the need for costly updates of drinking water systems throughout the country.

In Washington, D.C., for example, lead contamination discovered in the drinking water in 2004 prompted a $400 million plan to replace tens of thousands of service pipes made of lead. "Of approximately 130,000 residences served by the District of Columbia Water and Sewer Authority (DCWASA), an estimated 18 percent have lead service pipes," according to a report in *Environmental Health Perspectives,* the peer-reviewed journal of the National Institute of Environmental Health Sciences, part of the National Institutes of Health.[11] "Lead is in some older solder and plumbing fixtures as well."

The article goes on to point out that an "average 10 percent to 20 percent of U.S. environmental lead exposure comes from drinking water."

The dangers of lead. According to the EPA, homes built before 1986 are more likely to have lead pipes, fixtures, and solder. The following information about the dangers of lead appears on the EPA web site.[12]

Complications associated with lead poisoning include:

- Reduced IQ
- Slowed body growth
- Hearing problems
- Behavior or attention problems
- Failure at school
- Kidney damage

The symptoms of lead poisoning may include:

- Irritability
- Aggressive behavior
- Low appetite and energy
- Difficulty sleeping
- Headaches
- Reduced sensations
- Loss of previous developmental skills (in young children)
- Anemia
- Constipation

- Abdominal pain and cramping (usually the first sign of a high, toxic dose of lead poison)
- Very high levels may cause vomiting, staggering gait, muscle weakness, seizures, or coma

To reduce your chances of lead poisoning:

- Run tap water before drinking or cooking with it.
- Use cold water for drinking or cooking.
- Never cook or mix infant formula using hot tap water.
- Do not consume water that has been in your home's plumbing for more than six hours.
- Certain faucets and pitcher filters will remove lead from drinking water; be sure yours is certified to do so by NSF International.

For more information:

- National Lead Information Center: http://epa.gov/lead/
- Safe Drinking Water Hotline: 1-800-426-4791 (more about the hotline at http://water.epa.gov/aboutow/ogwdw/hotline/index.cfm
- NSF International: http://nsf.org

Government Assistance

Regions, states, cities, municipalities, and more picked up a short jump-start on coping with some of their water-related stresses and strains in the form of funds from the American Recovery and Reinvestment Act of 2009.[13]

In April 2009, the U.S. Department of the Interior announced $945 million in projects to be funded with Recovery Act allotments. The projects included meeting future water supply needs like water reclamation and reuse and rural water projects; infrastructure reliability and safety; environmental/ecosystem restoration; green buildings; water conservation initiatives, and drought relief.

The U.S. Environmental Protection Agency also announced $6 billion in stimulus funding for various programs, including the Clean Water State Revolving Fund program, which received $4 billion, and the Drinking Water State Revolving Fund program, $2 billion, to be allocated to various programs across the country.

Unfortunately, together these allocations and projects represent only a tiny portion of what's ailing America's water infrastructure.

WATER OVERUSE AND WASTE

For years, water in the United States was taken for granted as an inexhaustible, self-replenishing resource. Who hasn't run the dishwasher or washing machine when it wasn't quite full, left the water running—inside or out—when it wasn't necessary, or dawdled in the shower? Keeping in mind that each individual in the United States uses, on average, 100 gallons of water a day, most people could save 30 percent of that (about 30 gallons a day) simply by using more efficient water fixtures and appliances and modifying their behaviors. That's the word from WaterSense, an EPA program to help Americans conserve water. WaterSense breaks down some of the water-wasting ways that add up to billions of gallons of water down the drain every year:[14]

- Americans waste nearly 640 billion gallons of water every year flushing old, inefficient toilets. That's equivalent to 15 days' worth of water flow over Niagara Falls.
- 5 percent to 10 percent of U.S. homes have water leaks that drip away 90 gallons or more of water a day. That's enough water to fill two big bathtubs.
- Allowing a faucet to run for just five minutes a day uses as much energy as leaving on a light with a 60-watt bulb for 14 hours.
- Turning off the tap while you brush your teeth can save eight gallons of water a day.
- Some homeowners use as much as 70 percent of their household water for outdoor irrigation, with 50 percent of that wasted due to evaporation, wind, or runoff from overwatering.
- A full bathtub can take 50 to 70 gallons of water, while a five-minute shower uses only 10 to 25 gallons of water.

WATER FACTS

Putting wasting water in perspective, one leaky faucet adds up. That one drip is a fraction of a milliliter, but in one day a dripping faucet translates to approximately 259,200 drips, or 17 gallons of water a day and 6,248 gallons a year, according to the USGS's Drip Accumulator (http://ga.water.usgs.gov/edu/sc4.html)!

Too Much Equals Too Little

Plenty of rainfall doesn't necessarily solve an area's water supply problems, either. Look at central Florida. In the mid-1970s, the state had tons of water but was running out at the same time, says architect and planner Daniel E. Williams, who participated in a study called "The Green Swamp" that dealt with water supplies in central Florida. The problem, says Williams, was (and still is) that despite the area's huge amount of precipitation, the aquifers were built over, and the swamps were drained to accommodate growth. These areas built themselves out of a sustainable water system. "There are places like Tampa, for example, that already are desalinating for potable water," says Williams. "This in a place that gets fifty to sixty inches of rainfall a year. It's just really bad planning and a misunderstanding of the water system that's put the area in a position (where) they're now looking for water."

In the 1970s, 1980s, and 1990s, the Tampa Bay area, including Hillsborough and Pinellas counties and several municipalities, including the cities of Tampa and St. Petersburg, looked to water well fields, primarily in northwest Hillsborough and Pasco counties, to supply the area's water. But a lot of negative environmental impact occurred with a number of those wells, says Roger W. Sims, water resources, environmental, and land-use law partner with Holland & Knight, in Orlando, Florida. The notorious Tampa Bay Water Wars ensued. A lot of money and time was wasted fighting over who had the right to the water, whose fault the problems were, and more. Ultimately, the bickering parties stopped fighting and formed the Tampa Bay Partnership. "That's how Tampa Bay Water was created and the emphasis was shifted from fighting over water to finding solutions and new supplies," says Sims. "The legal fight had raged on and on and on, and in the meantime, the well fields kept pumping."

Those new supplies Tampa Bay Water came up with include a reservoir on the Alafia River to divert surface water to storage for dry periods, and a seawater desalination plant on eastern Tampa Bay. However, it's far from back to water as usual for the Tampa Bay Water area. As mentioned earlier, in March 2009, Tampa Bay's 15 billion-gallon reservoir officially ran out of water: Its level fell too low to pump and the Southwest Florida Water Management District pronounced it "dry."[15] Cracks in the walls of the aboveground reservoir added to the problem. The Tampa Bay utility scrambled to use other sources to satisfy the area's water needs. The reservoir has since been replenished and tight restrictions on water use lifted, but will the lesson be learned? Time will tell.

WATERTIGHT LAND DEVELOPMENT

How many times have you driven by (or maneuvered through) a stretch of road or parking lot flooded with standing water during or after a rainstorm or snowmelt? It may not have been a very heavy storm, but the water annoyingly pooled up anyway because it had nowhere to go. Its escape route has been paved or cemented over.

The manner in which land traditionally has been developed actually interferes with the natural replenishment of our water supply. Even in water-rich areas, no matter how much precipitation falls from the sky, large expanses of paved parking lots, roads, walkways, and impervious roofs keep the water out. Rather than soaking back into the ground and replenishing aquifers, water pools and evaporates away or runs off into the sewer systems and is carried away—usually far away.

"When it comes to regional and urban design and site planning, we've spent years disregarding the necessary connection between urban patterns and their geo-hydrological relationship to our drinking water," says urban design sustainability expert Williams. "Instead we have created millions of square miles of impervious surfaces and untold lineal miles of drainage canals, and then look for additional water once the swamp has been drained."

"In a natural system, water is captured, stored, and released, which provides both flood mitigation and water supply needs. In an urban system, the traditional challenge has been one of getting the water out of the neighborhood as soon as possible—typically sending it underground or shunting it to the ocean or bay," says Williams. "Now that water has become the limit to growth and development, the challenge is how to store the water within the urban and regional pattern so that communities utilize the water that lands there."

Williams suggests that urban areas that historically flood could be designed with small parks that pond or store water, "creating a pattern of open spaces and neighborhood gardens paid for by storm water dollars—a win-win."

"We need to connect design and science—then you have real solutions integrated into the urban fabric," he adds.

Down the Drain—Literally

Remember the Boston metropolitan area's loss of potable water through sewer pipes? It's not just leaky pipes that collect and lose good water. "Add the impact of paved and impervious surfaces in metropolitan areas, urban centers, suburbs, and exurbs in that same forty-three-town metropolitan area, and in a dry year add another 50 billion gallons of water that goes into storm drains and is instantly lost rather than infiltrating back into the ground," says Zimmerman of the Charles River Watershed Association. "In a wet year, it's as much as 120 billion gallons. Added to the potable water losses in sewer pipes for those same forty-three towns, we're throwing away the equivalent of two Charles Rivers in their entirety every single year!"

Zimmerman bases his numbers on his organization's analysis as well as data from the 2002 study by American Rivers, the Natural Resources Defense Council, and Smart Growth America titled "Paving Our Way to Water Shortages: How Sprawl Aggravates the Effects of Drought."[16]

"With the exception of the semi-arid and arid Southwest, the notion that somehow urban America is running out of water is ridiculous," says Zimmerman. "Instead, we're throwing our water wealth away through sewer pipes and storm drains. Perpetuating this sort of gray-pipe infrastructure will condemn us to serious water shortages and exacerbate floods and pollution—the two phenomena are closely linked. We must, instead, mimic the way nature treats rainwater, allowing it to percolate into the ground and support plant life even in our most dense urban areas."

Nature Stymied

The American Rivers/Natural Resources Defense Council/Smart Growth America study looked at the 20 metropolitan areas that experienced the most suburban sprawl between 1982 and 1997. They focused on sprawl—with its accompanying broad expanses of impervious surfaces—to show how it encroaches on open space and threatens water supplies, in part because the loss of undeveloped land means a loss of the land's natural filtering ability, preventing the seepage of water into the ground to replenish aquifers. Instead, the rain runs off the nonporous surfaces and is swept far away. "When we sprawl, we threaten our freshwater resources at the very time our demand for them is increasing. The large number of hard surfaces created by traditional suburban development fundamentally alters the local movement and availability of water," the study says.

In 1997, the volume of water that was runoff as opposed to soaking back into the ground in Atlanta, Georgia, was enough water to supply the average daily household needs of 1.5 million to 3.6 million people per year.

Source: "Paving Our Way to Water Shortages: How Sprawl Aggravates the Effects of Drought," © American Rivers, the Natural Resources Defense Council, and Smart Growth America, 2002.

Whether suburban sprawl is the cause or not, the volume of that runoff is staggering. The potential amount of water not infiltrated back into the ground annually during the period 1982 to 1997 ranged from 6.2 billion to 14.4 billion gallons in Dallas to 56.9 billion to 132.8 billion gallons in Atlanta, the study estimates.

Here are more numbers on groundwater infiltration losses caused by the abundance of impervious surfaces in metropolitan areas, as calculated by the estimates:

- Boston, Massachusetts: 43.9 billion to 102.5 billion gallons
- Charlotte, North Carolina: 13.5 billion to 31.5 billion gallons
- Chicago, Illinois: 10.2 billion to 23.7 billion gallons
- Detroit, Michigan: 7.8 billion to 18.2 billion gallons
- Greensboro, North Carolina: 6.7 billion to 15.7 billion gallons
- Houston, Texas: 12.8 billion to 29.8 billion gallons
- Minneapolis–St. Paul, Minnesota: 9 billion to 21.1 billion gallons
- Nashville, Tennessee: 17.3 billion to 40.5 billion gallons
- Orlando, Florida: 9.2 billion to 21.5 billion gallons
- Philadelphia, Pennsylvania: 25.3 billion to 59 billion gallons
- Pittsburgh, Pennsylvania: 13.5 billion to 31.5 billion gallons
- Raleigh–Durham–Chapel Hill, North Carolina: 9.4 billion to 21.9 billion gallons
- Seattle, Washington: 10.5 billion to 24.6 billion gallons
- Tampa, Florida: 7.3 billion to 17 billion gallons
- Washington, DC: 23.8 billion to 55.6 billion gallons

Stopping the Flow

The good news is that it's possible to slow the flow of this suburban wash-away and its effect on our nation's water supplies. The American Rivers study advocates "smart growth" that includes varied infrastructure investment to allow for a better mix of open spaces and impervious ones, and efficient location of development that is mindful of reducing its impact on water supplies.

Similarly, the EPA promotes its Green Infrastructure initiative, which calls for practices that either use or mimic natural processes as one way to deal with storm water runoff in metropolitan areas across the country. That includes "green" roofs, porous pavement, landscaping, rain gardens, and vegetated swales, all of which help capture storm water runoff instead of allowing it to wash away into the sewers and beyond.

Going Green

Going green one barrel at a time is one way the federally established Interstate Commission on the Potomac River Basin (http://www.potomacriver.org/cms/) promotes better use of its water resources. The commission sells 60-gallon rain barrels and offers workshops on how to use them in your garden. On the other side of the country, King County, Washington (http://kingcounty.gov/), offers guidance for consumers on how to create your own rain barrel and cistern system. Many other organizations and government entities offer rain barrels and guidance on using them, too. You can check out a few at http://epa.gov/region3/p2/what-is-rainbarrel. pdf. The U.S. Environmental Protection Agency also explains step by step how to make your own (http://epa.gov/reg3esd1/garden/rainbarrel.html).

All kinds of water-savings initiatives have taken root in many American cities. Here are a few more ways they're working, according to the EPA:

- *Philadelphia, Pennsylvania:* Planting vegetated swales instead of paving berms along some roads to enhance recapture of storm water runoff.
- *Pittsburgh, Pennsylvania:* Investing millions of dollars to restore riparian corridors, install wetlands, and create wildlife habitats along Nine Mile Run, a highly degraded stream in the area.
- *Seattle, Washington:* Reducing the volume of storm runoff in some areas by as much as 80 percent by installing stepped, vegetated areas. The city also has adopted the Street Edge Alternative (SEA) design, which adds vegetated swales to reduce total potential surface runoff.
- *Portland, Oregon:* Vegetated curb areas along roads help decrease storm water runoff here, too. The city also is a leader in downspout disconnection as a way to reduce storm water runoff in sewers. The city pays homeowners $53 for

every downspout disconnected from the sewer system, which allows an estimated 1 billion gallons of water annually to infiltrate back into the ground.

Some of the nation's water districts and treatment facilities have adopted a system of storm water fees, too. The greater the impervious surface area, the higher the fees.

As with almost everything else to do with water, there are winners and losers, happy customers and not-so-happy ones.

In Chicago, meteorologist Amy Freeze has proposed another approach to dealing with storm water runoff. The storm water specialist, also chief meteorologist for Fox Chicago, is a proponent of a voluntary storm water overflow alert system. Freeze explains her Stormwater Action Alert Program (SAAP) as a voluntary initiative to help consumers cut down on water use when conditions are right for a heavy rain event that could overwhelm sewers, cause flooding, and ultimately degrade water quality. "The community-based initiative would be similar to the already successful national air quality initiative, Partners for Clean Air," says Freeze. "With water, if before the storm occurs, we can limit the volume of water in our sewers, it will go a long way toward improving the quality of our source water."

Freezes envisions Chicago as a starting point for the alert system, and then hopefully grow it in scope. In October 2010, Freeze and a number of Chicago-area experts sat down to discuss various solutions to combined sewer overflows and the possible role an alert system could play when conditions for overflow exist.

GROUNDWATER POLLUTION

Groundwater runoff isn't a concern only because of its volume and resulting reduction in underground water supplies. That same suburban runoff combines with urban runoff, mixes with various pollutants it picks up along the way, and creates one of the biggest sources of water and groundwater pollution in the country. Known as nonpoint source pollution (NSP), it affects drinking water supplies not only in the United States, but also around the world, because groundwater systems provide 25 to 40 percent of the world's drinking water.[17] Add man-made pollution, including agri-culture and irrigation runoff, to the naturally occurring kind, and the groundwater poisoning issue intensifies. According to the EPA, NPS is now recognized as the primary threat to water quality in the United States.[18]

Consider California, a state many see as an environmental leader. "There is no such thing as a clean, pristine aquifer system in California," says Richard Atwater, of Inland Empire Utilities Agency in Southern California. "Geologically, we have naturally occurring arsenic in the groundwater wells and a variety of man-made pol-lution, like nitrate pollution, from the legacy of agriculture or the aerospace industry in Southern California during World War II and the Vietnam War. We have solvents and different chemicals used by aerospace in rocket manufacturing. Every one of our groundwater basins or aquifers in Southern California either has a Superfund site or a significant number of contaminated wells."

We'll talk more in the next chapter about what comes out of your tap besides water. For now, let's just say that it can include everything from arsenic to selenium to pesticides, boron, molyb-denum, uranium, MTBE, methane, chromium-6, and much, much more. In March 2009, some homeowners in Fort Lupton, Colorado, had to cope with flammable water coming out of their faucets. Other locales around the country have experienced similar situa-tions. Some suspect the problems could be related to natural gas exploration, drilling, and wells in the areas that possibly contami-nate the private wells.

Natural Pollutants

In states like New Hampshire, New Jersey, and Maine, groundwater wells can be contaminated with too-high concentrations of naturally occurring arsenic.

Selenium is another contaminant found in groundwater in parts of California and South Dakota. A naturally occurring, nor-mally trace element common in certain types of rock, selenium is vital to humans and animals in small quantities but can be toxic in higher concentrations. Often the selenium concentrations are aggravated by irrigation runoff. That's what happened in the 1980s to the Kesterson Reservoir, which served California's San Joaquin Valley. After studies showed severe deformities in wildlife as a result of dangerous selenium levels, the reservoir was closed and buried in the 1980s.

In Waukesha and New Berlin, Wisconsin, near Milwaukee, the natural contaminant in some of the region's deep bedrock aquifers is radium, a colorless, odorless, tasteless element.

Man-Made Pollutants

Irrigation-induced groundwater contamination is a serious threat across much of the western United States. With its semiarid climate and extreme dependence on underground aquifers and groundwater for drinking water supplies, the West is especially susceptible to man-made pollution. Arizona, for example, gets 40 percent of its drinking water from underground aquifers. When that state's ecosystems are disrupted or altered in some way—more water is taken out than is replenished, or a portion becomes polluted for whatever reason—supplies of safe water shrink.

With the pollution problems at Kesterson Reservoir still fresh, the U.S. Department of the Interior initiated the National Irrigation Water Quality Program in 1985 to look at possible irrigation contamination of water supplies in 26 locations across the West. Subsequent analysis of the data turned up selenium contamination in 12 of those areas. Some sites that were found to be contaminated are Belle Fourche Reclamation Project, South Dakota; Dolores–Ute Mountain area, Colorado; Gunnison River Basin–Grand Valley Project, Colorado; Kendrick Reclamation Project, Wyoming; Middle Arkansas River Basin, Colorado and Kansas; Middle Green River Basin, Utah; Pine River area, Colorado; Riverton Reclamation Project, Wyoming; Salton Sea area, California; Sun River area, Montana; Tulare Lake Bed area, California; San Juan River area, New Mexico; and Vermejo Project area, New Mexico.[19]

On the East Coast, farmland runoff is a problem for Chesapeake Bay's water quality because of excess nutrients from farm fertilizers, says Philadelphia-based water expert Kenneth J. Warren, an attorney with Hangley Aronchick Segal & Pudlin. He's also general counsel for the Delaware River Basin Commission, which manages water resources for approximately 15 million people who count on the Delaware River and its tributaries for their water supply. Numerous streams, in addition to those that drain into the Chesapeake, contain excess nutrients, he adds.

Nonetheless, from a recent historical perspective, the overall water quality of U.S. rivers has improved. "There's no doubt we're far better off today than thirty years ago," says hydrologist Waldrop, who has been studying water and environmental quality issues since the early 1970s. For example, he recalls the now-infamous fire on the pollution-choked Cuyahoga River in Cleveland, Ohio, in 1969. Things like that used to be regular occurrences, he says, but today the really bad culprits have been cleaned up.

Waldrop agrees that the water quality issue that has not been addressed is the nonpoint source pollution from things like agricultural and urban runoff. "Most of your water treatment plants don't treat for the chemicals we spray on our fields, the medicines people flush down the toilets, or the antifreeze that leaks out of our cars."

People are beginning to realize, too, that airborne pollution contributes to water pollution. "If something goes up a [smoke] stack, it gets into the atmosphere and eventually comes back to Earth and ends up in the water," Waldrop adds. "We're finding traces of certain combustion products from things like power plants in water supplies, and they didn't get there by a discharge. It's definitely an airborne transport of pollution."

GEOGRAPHY, GEOLOGY, AND WATER LAWS

The sites in the water quality selenium study mentioned earlier were selected in part because of their climate and high probability of irrigation-related contamination. Geology, especially the type of rock involved, as well as topography, also figured into the picture.

Geography and geology affect the amount as well as the purity of water supplies in many areas. It's not only a matter of keeping underground aquifers brimming. As occurs with selenium contamination, certain rock is more conducive to creating ground-water pollution. Arsenic, for example, can be prevalent and contaminate drinking water supplies in areas with basalt and volcanic rock. New Mexico is one example of that, says Sandia's Hightower.

Laws governing who gets how much water and from where complicate the mix. If one city's well or aquifer is polluted, that city probably can't just drill deeper or go somewhere else to satisfy its water needs.

Water-Rich Midwest Thirsts

Location, geology, and international law add up to water shortages in small New Berlin, Wisconsin, even though the town isn't far from Lake Michigan, part of the Great Lakes Basin, the largest freshwater lake system on the planet. The town already gets part of its water from the city of Milwaukee, which taps Lake Michigan. The lake borders Canada, so its water is regulated by international law.

Exacerbating New Berlin's problem is the fact the city is split by a subcontinental divide. The east side of the city (and of the divide) drains off the land into the Great Lakes Basin; on the other side of the divide and the city, water flows away from the lake into the Mississippi River Basin. In the scheme of access to water and international law, that makes a big difference. Water law, which we'll talk more about in Chapter 5, can involve the premise that what you take out of one basin or watershed has to go back to the same basin. That doesn't always happen in New Berlin (or elsewhere) because its drainage pattern prevents all the water returning to or replenishing its source.

New Berlin's water shortage is expected to reach 3 million gallons of water a day by 2020. And that's for a city of only about 40,000 people. To get more water—in this case, from Lake Michigan and the Great Lakes Basin—New Berlin petitioned the Wisconsin Department of Natural Resources for a diversion of Great Lakes water. Complicating matters, Wisconsin's use of Great Lakes water is affected by international law and by the

Great Lakes Basin Compact, designed to protect and conserve Great Lakes water. The compact does allow for diversion to straddling communities, but imposes certain restrictions. This was the first case in which an entity straddling the Great Lakes Basin applied for a diversion of Great Lakes water under the terms of the recently enacted compact, says Eric Ebersberger, water use section chief for the Wisconsin Department of Natural Resources. On May 21, 2009, New Berlin's request for additional water was granted. However, part of the approval included a stipulation that the city cut by 10 percent its per-capita residential water use by 2020.[20]

POPULATION LOCATION AND GROWTH

Cities, towns, and municipalities across the country face shrinking supplies of water to meet growing demands. Unfortunately, population sprawl hasn't paid much attention to water supplies.

"The issue with water within ten to twenty years may rival what we have with energy in the sense that right now it's getting hard to find, and like oil [water] will become much more expensive," says hydrometeorologist Frank Richards. "The fundamental way forward is to understand the problem of water scarcity, and through a societal approach, deal with it."

For example, if enough water isn't available to meet the potential needs of a proposed project, the area won't be developed. Theoretically, says Richards, that's what's happening in Las Vegas, where water is always an issue. He points to zoning codes that require developers to identify the water supply for any potential development before the project is approved. "Developers will go into central Nevada to try to identify an aquifer they can draw from. That's an onerous burden on developers, but frankly, in some areas, it's the only way we can mitigate the issues," Richards says.

In pioneer times, settlements tended to spring up around water supplies. These days, it doesn't seem to matter. The vast majority of people in the United States take water for granted, and access to water generally doesn't top their list of concerns. That's likely to change, experts agree, as water takes center stage in the twenty-first century.

WATER REALITIES

- Climates have always been changing. But add in global warming, and it can create a crisis for the nation's water supplies.
- In addition to global warming and climate change, culprits that squeeze the nation's water supplies include changes in population location and size; pollution, both natural and man-made; water and land use and overuse; strict water-use regulations or lack thereof; antiquated or inefficient and worn-out infrastructure; outdated water treatment plants that can't handle twenty-first-century poisons; water waste and what to do with it; and methods of developing land that prevent natural replenishment of groundwater.
- Antiquated infrastructure represents a tremendous threat to our water supply. As amazing as it seems, more than half of the wastewater that runs off and away via many cities' metropolitan wastewater infrastructure actually is potable water that drains or leaks into the systems and then is carried away forever.
- "Fixing" the infrastructure problem will cost hundreds of billions of dollars.
- That leaky faucet at home adds up to plenty of water waste— one drip could send more than 6,000 gallons of water down the drain over a year's time.
- Even when there seems to be enough water, other issues— including geological and topographical features such as drainage basins and continental and subcontinental divides—can lead to short supplies and even to water rationing.
- Not all water pollution is man-made. Arsenic contamination, for example, can occur naturally in areas with basalt and volcanic rock.

NOTES

1. FAO, "Comprehensive Assessment of Water in Agriculture, 2007: Areas of Physical and Economic Scarcity," http://www.fao.org/nr/water/art/2007/scarcity.html.
2. National Resources Defense Council and Tetra Tech, "Climate Change, Water, and Risk: Current Water Demands Are Not Sustainable" (July 2010), http://nrdc.org/globalWarming/watersustainability/.
3. National Oceanic and Atmospheric Administration, National Climatic Data Center, "Climate of 2007 Annual Review U.S. Drought," http://www.ncdc.noaa.gov/oa/climate/research/2007/ann/drought-summary.html.
4. U.S. Environmental Protection Agency, "Possible Water Resource Impacts in North America," http://epa.gov/climatechange/effects/water/northamerica.html#ref.
5. U.S. Global Change Research Program, "Global Climate Change Impacts in the United States," June 16, 2009, http://globalchange.gov/publications/reports/scientific-assessments/us-impacts/full-report.
6. Southern Nevada Water Authority, "Valley Working to Keep Drought at Bay," http://www.snwa.com/html/drought_index.html.
7. American Society of Civil Engineers, press release, "Report Card for American Infrastructure: Drinking Water" (2009), http://www.infrastructurereportcard.org/fact-sheet/drinking-water/.
8. Massachusetts Water Resources Authority, "Infiltration/Inflow Task Force Report: A Guidance Document for MWRA Member Sewer Communities and Regional Stakeholders" (March 2001), 5.
9. U.S. General Accounting Office, GAO Highlights (GAO-04-461; March 2004), "Water Infrastructure: Comprehensive Asset Management Has Potential to Help Utilities Better Identify Needs and Plan Future Investments," 13, http://gao.gov/new.items/d04461.pdf.
10. American Water Works Association, Water Industry Technical Action Fund, "Dawn of the Replacement Era: Reinvesting in Drinking Water Infrastructure" (May 2001), http://win-water.org/reports/infrastructure.pdf.
11. Tina Adler, "Lead: Washington's Water Woes," National Institute of Environmental Health Sciences, National Institutes of Health, http://www.pubmedcentral.nih.gov/articlerender.fcgi?artid=1247550.
12. U.S. Environmental Protection Agency, "Lead in Drinking Water," http://epa.gov.
13. Department of the Interior Recovery Investments, press release, "Summary of Projects: American Recovery and Reinvestment Act," April 15, 2009, http://recovery.doi.gov/bor/summary_projects.php; U.S. Environmental Protection Agency, press release, "Clean Water and Drinking Water State Revolving Funds," http://epa.gov/ow/eparecovery/.
14. WaterSense, U.S. Environmental Protection Agency, http://epa.gov/watersense/.
15. "Tampa Bay Water Service Area under Critical Water Shortage Alert," Press release, March 31, 2009, www.swfwmd.state.fl.us/news/article/1219/.

16. Betsy Otto, Katherine Ransel, Jason Todd, Deron Lovaas, Hannah Stutzman, and John Bailey, "Paving Our Way to Water Shortages: How Sprawl Aggravates the Effects of Drought" (2002), American Rivers, the Natural Resources Defense Council, and Smart Growth America, http://smartgrowthamerica.org/ DroughtSprawlReport09.pdf; http://www.nrdc.org/media/docs/020828.pdf.

17. U.N. Educational, Scientific and Cultural Organization, "Facts and Figures extracted from the 2nd United Nations World Water Development Report," www.unesco.org/water/wwap/wwdr/wwdr2/facts_figures/index.shtml.

18. U.S. Environmental Protection Agency, "Green Infrastructure," http://cfpub .epa.gov/npdes/greeninfrastructure/technology.cfm.

19. Ralph L. Seiler, Joseph P. Skorupa, David L. Naftz, and B. Thomas Nolan, "Irrigation-Induced Contamination of Water, Sediment, and Biota in the Western United States—Synthesis of Data from the National Irrigation Water Quality Program" (November 2003), p. 50, U.S. Geological Survey (USGS Professional Paper 1655), http://pubs.usgs.gov/pp/pp1655/pp1655_v1.1.pdf; National Irrigation Water Quality Program, U.S. Department of the Interior, "Review of the Department of the Interior's National Irrigation Water Quality Program: Planning and Remediation," Committee on Planning and Remediation for Irrigation-Induced Water Quality Problems, Water Science and Technology Board, Commission on Geosciences, Environment, and Resources, National Research Council, http://www.usbr.gov/niwqp/abstracts/Committee2.doc.

20. State of Wisconsin Department of Natural Resources Water Division, "Water Supply Service Area Plan & Diversion Approval," May 21, 2009, http://www.dnr.state.wi.us/org/water/dwg/greatlakes/NBerlin/NBerlin_ WaterSupplySAP_final.pdf.

4

DANGER!
SAFE WATER AT RISK

*Understanding where your water comes from and what people are
trying to do to protect these water resources is something we all can do.*
—Greg Delzer, national coordinator
of source water-quality assessments,
U.S. Geological survey

Today, having plenty of water does not ensure survival. In some
countries, survival becomes questionable not because of the lack of
water but because the water supply can carry deadly toxins.

Fortunately, America is one of the few places in the world where
100 percent of the population has access to safe drinking water—or
so we've all thought. The reality, however, is that even here thousands of people get sick every year as a result of the water they drink.

It's generally hush-hush, but people are poisoned in a variety of
ways and degrees by the very water that comes from their wells and
water treatment facilities. The nonprofit Natural Resources Defense
Council (NRDC) estimates that more than 7 million Americans get
sick from contaminated water every year. Those numbers may be
high or low, depending on who is keeping track and whether water-
related illnesses are reported. Many aren't, according to the U.S.
Environmental Protection Agency (EPA).[1]

Officially, the EPA, the U.S. Centers for Disease Control and Prevention (CDC), and the Council of State and Territorial Epidemiologists collaborate to collect information on waterborne disease outbreaks—what they refer to as WBDO—across the country. According to a 2006 report from the group, between 1991 and 2002, "an estimated 403,000 persons became ill, 4,400 persons were hospitalized, and 50 persons died" as a result of waterborne diseases. Among the causes of WBDOs that the report cites are water distribution systems' deficiency, untreated groundwater, deficiency in water treatment, and "unknown."[2]

TWENTY-FIRST CENTURY REALITY

Whatever may be the cause of reported and unreported illnesses from water supplies, one thing is clear. Those antiquated water delivery systems and outdated water treatment plants mentioned in Chapter 3 contribute to more ills than wasted water down the drain. Aging pipes (see Figure 4.1) that break can allow contaminants that breed bacteria into the water. Out-of-date water treatment facilities may combat most of the parasites and bacteria, but they don't necessarily work with twenty-first-century contaminants like pesticides, industrial chemicals, and pharmaceuticals. Both man-made and naturally occurring contaminants can end up in source water, too. (*Source water* refers to the origin of the water that comes out of your tap; it could be a shallow or deep aquifer, a river, a lake, an ocean, or any or all of these.)

What Can Contaminate Water Sources?

The EPA suggests some of the ways in which water sources like aquifers, lakes, rivers, and streams can be contaminated.[3]

- *Microbial contaminants:* These include bacteria and viruses that can come from animal waste, sewage treatment plants, and septic systems.
- *Inorganic contaminants:* These include metals and salts and can occur naturally or from storm water runoff in urban areas, wastewater discharge, oil and gas production, mining, or farming.

FIGURE 4.1 A Portion of a Clogged Water Pipe from Rural Iowa
Source: Iowa Rural Water Association.

- *Pesticides and herbicides:* These come from agriculture, irrigation, and storm water runoff as well as residential runoff.
- *Organic chemical contaminants:* These include synthetic and volatile organic chemicals, which are by-products of industrial processes and petroleum production and can also come from gas stations, urban storm water runoff, and septic systems.
- *Radioactive materials:* These can occur naturally or as a result of oil and gas production or mining activities.

Chemicals in Source Water

Low levels of man-made chemicals turned up in a number of source water and public water supplies before and after treatment, according to a report released in December 2008 by the National Water-Quality Assessment (NAWQA) Program of the U.S. Geological Survey (USGS).[4]

For the study, scientists tested water in nine rivers, all of which are sources of public water systems, for the presence of 260 common chemicals ranging from pesticides and solvents to gasoline hydrocarbons, manufacturing additives, disinfection by-products, and more. Low concentrations of 130 of those chemicals were found in streams and rivers before they entered public drinking water treatment plants. Almost two-thirds of the chemicals also were in the water *after* treatment. The water treatment technology and processes did not remove the contaminants. The greater the proportion of agricultural and urban land in a watershed area, the greater the number of chemicals and the higher their concentration. More chemicals at higher concentrations also turned up when wastewater was discharged upstream from a community water treatment plant, according to the report. "Most of the chemicals were at levels equivalent to one thimble of water in an Olympic-sized pool," USGS officials said in releasing the report.[5]

"Low-level detection does not necessarily indicate a concern to human health, but rather indicates what types of chemicals we can expect to find in different areas of the country," adds the USGS report's lead scientist, Gregory Delzer, national coordinator of source water-quality assessments, USGS. "What was surprising was that conventional water treatment *did* in fact remove a number of these compounds. Another surprising finding was that because conventional water treatment was not specifically designed to remove those contaminants, mixtures became pretty evident—75 percent of all our samples had at least five of the compounds present and 50 percent of the compounds had about fourteen of these compounds present in every sample," said Delzer.

Scientists don't know the actual effects of the chemicals in low concentrations on humans. However, the study points out that "the common occurrence of chemical mixtures means that the total combined toxicity may be greater than that of any single contaminant present."

Among the rivers tested for chemicals were:

- White River (Indiana)
- Elm Fork Trinity River (Texas)
- Potomac River (Maryland)
- Neuse River (North Carolina)
- Chattahoochee River (Georgia)

- Running Gutter Brook (Massachusetts)
- Clackamas River (Oregon)
- Truckee River (Nevada)
- Cache la Poudre (Colorado)

When and Where You Least Expect It

In 2003, another study, titled "What's on Tap? Grading Drinking Water in U.S. Cities," by the nonprofit Natural Resources Defense Council (NRDC), looked at the quality of drinking water systems in 19 U.S. cities and found much more than water in the water.[6]

Here are some of the unwanted (and sometimes hazardous) extras they discovered in the water.

- *Rocket fuel:* Perchlorate, harmful to the thyroid and possibly carcinogenic, is in the water of 20 million Americans. High levels have been measured at times in Los Angeles, Phoenix, and San Diego.
- *Lead:* This substance, which can cause brain damage and decreased intelligence in children, gets into drinking water via corroding pipes and faucets. Boston, Newark, and Seattle exceeded the national action level for lead.
- *Germs:* These included coliform bacteria and *Cryptosporidium,* a microscopic disease-carrying protozoan. The Natural Resources Defense Council's study found that many cities should be concerned about their water supply's vulnerability to such contamination.
- *Arsenic:* Recently judged not safe at any level in drinking water, it's still present at significant levels in the drinking water of 22 million Americans.
- *Contaminant levels:* Spikes are on the rise, a sign that aging pipes and water-treatment facilities often can't handle today's contaminant loads. (Such loads may occur, for example, after a major storm or an industrial spill.) In recent years, Atlanta, Baltimore, and Washington, DC, all issued boil-water alerts in response to such spikes.
- *Groundwater supplies:* These can also be vulnerable to contamination. Fresno, California's groundwater is becoming seriously

compromised by agricultural and industrial pollution, including nitrates; Albuquerque, New Mexico's groundwater is over-taxed and threatened by pollutants from numerous sources.

That study concluded that pollution, deteriorating waterworks, and out-of-date water-treatment technology sometimes led to health risks. Only one city's water, Chicago's, was rated "excellent." The study isn't environmentalist banter or propaganda. It was peer-reviewed for accuracy and impartiality by independent experts.

Water contamination doesn't happen only on a citywide scale, and sometimes the perpetrators don't even realize it's occurring. It happens to individual homeowners, too, with what can be devastating results. Steve McIver found that out firsthand when he and his wife wanted to sell their West Deptford Township, New Jersey, home. As is customary in their area, where many homes have furnaces that use oil stored in underground tanks, they had to have their oil tank certified as sound before they could sell the home. However, the state inspector wouldn't certify the tank after soil tests hinted at a problem. When the tank was removed, it turned out to have a "small" leak, says McIver. "I never saw more oil than somewhere between the size of a fist and a grapefruit. But they dug dirt out by the dump-truck full."

Fortunately for McIver's pocketbook, but not so fortunately for the state of New Jersey, the oil had reached the water table. The water is owned by the state of New Jersey, so McIver's homeowner's insurance handled the roughly $45,000 cleanup as a liability claim. "If it hadn't reached the water table, we would have had to pay for all of it out of pocket," McIver says. "I know one person who had a similar situation with a heating oil tank, but the oil hadn't reached the water table. He had to pay $100,000 out of pocket and ended up declaring bankruptcy."

Less than Pure Tap Water

An Associated Press investigation also found more than water coming out of the tap in a number of cities across the country. In a 2008 study, the news organization found traces of pharmaceuticals including sex hormones, antibiotics, anticonvulsants, mood stabilizers, and more in the drinking water supplies of at least 46 million Americans. Admittedly, the amount of the drugs was listed as "trace," but how much is too much? What if you're deathly allergic to one of those drugs and a "trace" of it slips into your water supply? What are the ramifications for you and your children? That trace could end up being a fatal dose.

Consumers also often unwittingly contribute to water contamination, tossing outdated or unwanted pharmaceuticals down the drain. That's one way trace amounts of drugs get into water supplies. To help deal with the problem, some organizations, police, and government agencies now sponsor collection sites or events for outdated, unwanted, or unused drugs. Turning those drugs in instead of tossing them out helps keep our water supplies cleaner.

Other studies by other reputable news organizations, including the *New York Times,* single out other equally concerning contaminants in the water coming out of our water taps. The U.S. EPA has a web site that details various drinking water contaminants and regulations related to them (http://water.epa.gov/drink/contaminants/index.cfm#List).

The Feds Are Concerned

In the wake of these and other reports of contaminants in our water supplies, federal and state agencies and Congress now look more closely at exactly what's in the water. In testimony before the Water Resources Subcommittee of the U.S. House of Representatives on September 18, 2008, U.S. Representative Carolyn McCarthy (D-New York) talked about what and how pharmaceuticals end up in water supplies:

> In my state of New York, health officials found heart medicine, infection fighters, estrogen, mood stabilizers, and a tranquilizer in the upstate water supply. Six pharmaceuticals were found in the drinking water right here in Washington, DC. We don't know how the pharmaceuticals enter the water supply. However, it is likely that some medications not fully absorbed by the body may have passed into the water through human waste. In other cases, unused pills may have simply been flushed down the toilet.

Additionally, some agricultural products and medications may have run off into groundwater supplies.

In addition to antibiotics and steroids, the EPA has identified over one hundred individual pharmaceuticals and personal care products (PPCPs) in environmental samples and drinking water. Wastewater treatment plants appear to be unable to remove pharmaceuticals from the water completely. The presence of the pharmaceuticals in our nation's waters raises serious questions about the effects on human health and wildlife.

Currently the EPA, working with public input, is looking at how to what it calls "reinvigorate" its approach to clean water. (Check out developments in EPA's ". . . Strategy for Clean Water" at http://blog.epa.gov/waterforum/.) Congress also continues to work to toughen rules related to the Clean Water Act, and to pass new laws that address some of the country's pressing water issues. In addition, as discussed earlier, the American Recovery and Reinvestment Act of 2009, signed into law by President Obama on February 17, 2009, allocates billions of dollars for water-related environmental, cleanup, and infrastructure needs.

WHAT YOU DON'T KNOW CAN HURT YOU

Meanwhile, health threats—real and unknown—continue to infiltrate the water that Americans across the country drink.

Not All Contaminants Are Man-Made

People, business, and industry aren't solely to blame for water pollution and contamination. Some of the most deadly contaminants come from Mother Nature, too. What's in water supplies often depends on what's in the surrounding rocks. As water is sucked out of certain rocks, naturally occurring minerals dissolved in the water are pulled out, too, and not all those minerals are good for you. In addition to arsenic, selenium, and radium mentioned previously, other naturally occurring minerals like uranium and various salts can seep into underground water supplies, especially in aquifers that have been drawn down significantly. The Arkansas River, which crosses parts of the mineral-rich Rocky Mountains in

Colorado, then flows into Kansas, Oklahoma, and Arkansas, is one river with some of these issues. The high concentrations of these contaminants in the Arkansas River in eastern Colorado and southwestern Kansas are mainly the result of agricultural consumption of the water, which leaves the salts behind. To illustrate how that happens, hydrogeologist Don Whittemore of the Kansas Geological Survey suggests the following analogy.

"Imagine a pot of water boiling on the stove. If you allow the pot to boil down until it's dry, a white scum is what's left. That white stuff is the salts in the water. It's much the same with the Arkansas River water diverted for irrigation. As the water evaporates and is consumed by crops, what remains are salts and minerals in more concentrated form. The saline water can then seep into and contaminate the underlying aquifer."

At least two cities along the Arkansas River in Kansas have recorded uranium levels well above the standard, says Whittemore, who is studying uranium levels along the Arkansas through a grant project. The city of Lakin already has decided to spend several million dollars on a new water treatment system to help deal with the problem.

Deadly Salmonella

Another consideration is the dreaded scenario of the science-fiction realm—a town's water supply mysteriously and suddenly poisoned. Imagine waking up one day to find out your tap water is off limits. In spring 2008, that's exactly what happened in small Alamosa, Colorado, after more than 300 people there suddenly came down with salmonella poisoning linked to their water supply. The town's nearly 8,000 residents were warned to drink bottled water only and not even to shower with the town's water for two weeks. The entire system needed to be flushed with chlorine to kill the bacteria, and it was nearly a month before Alamosa got its water back. It took more than a year for the Colorado Department of Public Health and Environment to issue a report on the cause of the contamination: "animal source of fecal contamination." Apparently the contaminants found their way into the city's above-ground storage reservoir, and then subsequently into the city's water system likely by way of "cracks and holes" in the reservoir walls (there's that old and antiquated infrastructure issue again). Further compounding the situation, Alamosa at the time did not chlorinate (disinfect) its water supply. That's changed![7]

For more on salmonella, check out the U.S. Centers for Disease Control and Prevention website (http://cdc.gov/salmonella/).

MTBE Risks

Across the country—especially along the East Coast—shallow groundwater supplies risk contamination by methyl tert-butyl ether (MTBE), a component used to add oxygen to gasoline. You can't taste MTBE, but research has shown possible human health consequences, including cancer and effects on reproduction and development. Not only are heavily urbanized areas at risk, but rural ones are, too.

One of the worst contaminations is in New Hampshire. Specifically, it is in the four counties where MTBE as a gasoline additive was mandated by law until January 2007—Rockingham, Strafford, Hillsborough, and Merrimack counties—according to a January 2008 USGS report.[8]

"In the four counties using reformulated gasoline, we found MTBE at or above 0.2 parts per billion (ppb) in 30 percent of public supply wells and in 17 percent of the private homeowner wells," USGS hydrologist and the study's lead author, Joseph Ayotte, said in issuing the study. "One in every three wells tested in Rockingham County had MTBE. In the most densely populated areas of the county, one in two wells tested contained MTBE. We also found that more than 70 percent of the water tested from wells serving mobile home communities in the state had MTBE. While levels are mostly very low, the study shows that MTBE occurs in groundwater throughout the state," he added.

The study goes on to point out that although MTBE no longer is used as a gasoline additive in the state, it's not known how long the chemical remains in the groundwater.

Arsenic, Too

If the man-made toxin MTBE doesn't get in the water, perhaps arsenic will. New Hampshire, along with Maine and other parts of New England, must deal with that naturally occurring unwelcome additive to some of their water supplies. (Arsenic can be an issue elsewhere, too—as far west as California.) The state of New Jersey even offers free water-testing kits to enable people to test for arsenic.

Some possible symptoms of arsenic poisoning, according to the U.S. Environmental Protection Agency, include the following:

- Thickening and discoloration of the skin
- Stomach pain, nausea, vomiting
- Diarrhea
- Numbness in hands and feet
- Partial paralysis
- Blindness

Arsenic also has been linked to cancer of the bladder, lungs, skin, kidney, nasal passages, liver, and prostate.

For more information, check out the EPA's arsenic information pages at http://epa.gov/safewater/arsenic/index.html.

In certain sediments and under certain conditions, arsenic can dissolve over time into groundwater supplies and can be especially prevalent in bedrock aquifer wells, also known as deep or artesian wells. This conclusion appears in a study by the USGS, National Cancer Institute (National Institutes of Health), Dartmouth Medical School, and the state departments of health in Maine, New Hampshire, and Vermont. That study,[9] released in 2006, cites geology as the primary factor related to arsenic in groundwater wells. "Arsenic in groundwater used for private or a public water supply is a significant public health concern," said Robert W. Varney, then regional administrator of the EPA's New England office. "To protect families, the EPA recommends that private well owners routinely test their drinking water for arsenic."

A NO-WIN SITUATION

In 2007, Los Angeles announced that it would have to drain two of its reservoirs because of bromate contamination. The bromate formed when water was treated with chlorine and exposed to sunlight. Chlorine is added to most drinking water at treatment plants.

Camp Lejeune Tragedy

Unfortunately, not all Americans get the word that their water could be dangerous and even deadly until too late. Tens of thousands of U.S. Marines, former Marines, and their families were exposed to contaminated drinking water, not in far-off realms, but at home. This happened at the Tarawa Terrace family housing complex at Camp Lejeune, North Carolina, from November 1957 through February 1987. The drinking water was contaminated with the dry-cleaning solvent PCE (tetrachloroethylene), a possible carcinogen that may cause cancer and birth defects. Although the military base shut down the Tarawa Terrace water treatment plant in 1987 "because of PCE contamination of the drinking water," government studies of possible ill effects of the water on the Marines and their families are relatively recent. In June 2007, federal health investigators reported:

> A new analysis shows that former Marines and their families who lived in Tarawa Terrace family housing units during the period November 1957 through February 1987 received contaminated drinking water containing the dry-cleaning solvent tetrachloroethylene (PCE). Levels of PCE in the drinking water during this period exceeded the amount currently allowed by the Environmental Protection Agency under the Safe Drinking Water Act. Exposure to PCE-contaminated drinking water occurred because PCE leaked into groundwater that supplied the Tarawa Terrace drinking water system from a dry cleaner located outside the Camp Lejeune military base.

Today, the Agency for Toxic Substances and Disease Registry, an agency of the U.S. Department of Health and Human Services maintains a Camp Lejeune information site on their web site. (http://www.atsdr.cdc.gov/sites/lejeune/events.html).[10]

The Agriculture Factor in the Pollution Equation

Agricultural irrigation, as we mentioned earlier in this chapter, can contribute to groundwater pollution problems. As excess water runs off fields and into rivers and streams or seeps into aquifers, it often carries with it fertilizer and chemical residues. It's similar to urban storm water runoff, with its own unique pollutants.

The Mississippi River and its problems with pollutants are a good example, says Tennessee hydrologist William Waldrop. Excess fertilizers and pesticides run off into the river and end up in the Gulf of Mexico, where there's a recurring dead zone because the pollutants deplete the oxygen. The mighty Mississippi River is a muddy mess these days, especially by the time it finishes meandering more than two thousand miles from its source in Minnesota, past New Orleans, and out into the Gulf of Mexico. Along the way, millions of people in towns, cities, and rural areas rely on it as the source of their drinking water, and as the dump for their wastewater. Experts agree that this makes for a rough combination.

Florida's troubled Lake Okeechobee, in the south-central portion of that state, is another example of agricultural pollution. Because this shallow lake is affected by drought, the state and its farmers have struggled for years with pollution in the form of high nitrogen and phosphorous levels from farming runoff. A battle brews there. One group wants to back-pump agriculture runoff into the lake to boost its water levels and keep farmers from financial meltdown, while another group—environmentalists—says that would be a big mistake.

On the East Coast, farmland runoff is a problem for Chesapeake Bay's water quality, too, in the form of excess nutrients from fertilizers used on farms, says Philadelphia-based water attorney and expert Kenneth J. Warren. He's also general counsel to the Delaware River Basin Commission, which manages water resources for approximately 15 million people who count on the Delaware River and its tributaries for their water supply. "There are also numerous [other] streams, in addition to those that drain into the Chesapeake, that contain excess nutrients," he adds.

Agricultural irrigation runoff also taints portions of the Colorado River. In this case, though, the culprit is the natural leaching of salts from the soil. The result, experts agree, is that the now high-saline-content water is of little use to farmers downstream.

WASTES THAT CAN CONTAMINATE GROUNDWATER

Michigan's Environmental Science and Services Division suggests possible ways that businesses contribute to groundwater contamination:*

- All businesses: Sanitary sewage
- Vehicle service/body repair: Oil, antifreeze, solvents, fuels, paints, metal residues
- Car/truck washing: Road salt, gasoline, antifreeze, oil-laden wash and rinse waters, cleaners
- Metal parts cleaning: Alkaline solutions, solvents, phosphate solutions, metal residues, rinse waters, oil and greases
- Laundromat operations: Dirty wash water, detergents, laundry pre-wash solvents
- Dry cleaning: Solvents, filters
- Furniture repair and refinishing: Solvents, paints, varnishes, shellac
- Photo finishing/silk screening/printing: Processing chemicals, inks
- Paint mixing: Paints, solvents, pigments
- Food processing: Food scraps and juices, wash water, cooling water, salt

*Michigan Environmental Science and Services Division, "Preventing Groundwater Contamination," http://michigan.gov/documents/deq/deq-ead-tas-grwtrcon_329692_7.pdf.

WHAT ABOUT YOUR WATER SUPPLY?

All this talk of water pollution and contamination isn't a scare tactic. It's reality. Should you worry about the safety of your water? That depends. When was your city/county/state water system built? Is it equipped to handle modern problems? If you rely on a private well, when was the last time it was tested for contaminants? These are questions you may want to ask yourself and the provider of your drinking water. As we've discussed, the answers may shock and dismay you.

To find out how safe your water source is, check out the EPA's local drinking water information web site, http://water.epa.gov/drink/local/, and follow the geographic links to your state and region. You can check out about protecting the source of your water at http://water.epa.gov/infrastructure/drinkingwater/sourcewater/protection/index.cfm.

PROTECTING WATER SUPPLIES

Many of the nation's cities, towns, water suppliers, industries, and private organizations already have begun to ask these crucial questions and continue to work on getting the answers right.

New York in Action

New York City's aging water system dates back to the early 1900s and earlier. The city and the federal government are spending $12 million to set up a contaminated-water warning system, and the city has spent another $300 million. New York City has the largest unfiltered water supply in the world. It depends on a series of reservoirs, the most distant of which is more than 100 miles north of the city. The reason for concern, though, isn't terrorists poisoning the water supply; it's fear that construction around the watershed will pollute the reservoirs. To get a better idea of how that can happen, consider how a watershed operates. Think of a swimming pool that's covered with a tarp or some other flexible cover. After a rainstorm, the water drains toward, and collects, in the center. That's what happens with a watershed—all points drain into the lowest point, which in this case is the reservoir or reservoirs (Figure 4.2).

How to Help Lessen Groundwater Contamination

Here are suggestions on how businesses can lessen the chance of groundwater contamination, according to the Michigan Environmental Services Division.[11]

- *Practice waste reduction.* List the waste your business generates currently, and then figure out if you can reduce the toxicity and amount of that waste.
- *Replace toxic raw materials* with nontoxic or less-toxic raw materials wherever possible.
- *Replace toxic operational supplies* like cleaners and solvents with nontoxic or less-toxic materials wherever possible.

FIGURE 4.2   After a Rainstorm, Water Drains toward the Lowest Point
Source: U.S. Geological Survey.

- *Improve production process efficiency* so fewer raw materials end up as waste that must be disposed of. Reuse samples in production.
- *Encourage employees to think "waste reduction."* Better yet, include them in the planning process and give them an economic incentive to reduce waste.
- *Reuse and recycle process by-products as raw materials,* either on-site or in another company's process. Examples include process and cleaning solutions, wash water, rinse water, and cooling water.
- *Implement a computerized inventory control* system that helps ensure that your business limits chemical purchases to required amounts and provides alerts for possible spoilage or obsolescence.
- *Implement a computerized waste control system.* Make sure that wastes have proper labels so there is no need for future testing and analysis. Also, make sure that wastes are mixed only when there is no potential for reuse or recycling and when mixing is not prohibited by regulations.

- *Design process equipment to minimize waste.* Use a centrifuge rather than a filter, for example, to avoid filter cartridge disposal.
- *Implement an aggressive equipment maintenance program* to prevent leaks. Periodically check tanks, seals, pipe joints, and other equipment for needed repairs.
- *Contain and immediately clean up any spills, leaks, and drips.* Build or purchase secondary containment structures. Use drip pans under spigots and in other areas where there is likely to be seepage.
- *Use absorbent materials for spills when necessary.* Dispose of used absorbents with the same degree of care as you use with the materials being absorbed.
- *Install a catch basin in loading and unloading areas.* Nearly one-third of all accidental spills occur at loading docks. Keep rainwater and dirt out of the catch basin.
- *Hook up to a municipal sewer system if possible.* Pretreat process wastes to comply with local ordinances or applicable federal categorical pretreatment standards.
- *Route wastewater to holding tanks* that can be periodically pumped out. Do this if no municipal sewer system connection is available.
- *Hire a licensed liquid-industrial-waste hauler* to pick up the wastewater for proper disposal.
- *Perform all outside work on a sealed concrete or asphalt-paved surface* surrounded by a berm or dike.
- *Store raw materials and wastes under a roof or other protective cover* and on a sealed concrete or asphalt-paved surface. Provide additional secondary containment when necessary.
- *Make sure workers follow manufacturers' directions* when mixing materials to prevent using more materials than needed or making materials more hazardous than necessary. For large volumes and routine mixing, it is better to install an automated mixing system.
- *Do not dispose of items that contain hazardous materials* in trash that will be buried in a sanitary landfill or that will be incinerated. Recycle used fluorescent and high-intensity lamps, small batteries, capacitors containing polychlorinated biphenyls (PCBs), mercury thermometers, and other lab instruments, or handle them as hazardous waste.

- *Develop emergency response plans* that identify potential problem areas and address those actions necessary to reduce environmental and health risk. Under some circumstances, the law may require you to have these plans.
- *Learn more about groundwater and the impact you have on it.* Numerous resources are available through your library, by contacting environmental agencies, or on the Internet. For example, go to http://p2pays.org/ref/14/13673.htm or http://epa.gov/seahome/gwprimer.html.

Crackdown on Arsenic

Where you least expect it, arsenic surfaces. This time it's Cape Cod, Massachusetts, where arsenic turned up in aquifer sediments near a sewage treatment plant. Researchers from the U.S. Geological Survey have been studying how and why the arsenic was in the sediments even though the wastewater didn't have any appreciable concentrations.

Scientists found that the source of the arsenic was natural mineral coatings on the sediments of the sand and gravel aquifer through which the wastewater plume was moving. "The wastewater plume changed the chemistry of the aquifer, creating conditions where arsenic bound to the sediments and was released into the water," USGS reports. "The finding is significant because it demonstrates that naturally occurring arsenic, adsorbed onto the surfaces of quartz and other mineral grains, can be mobilized by human activities on the land surface with no link to arsenic. Land-use planners can use the results of this investigation to help identify and protect vulnerable underground drinking-water sources."[12]

BOTTLED INSTEAD?

Americans drank more than 8.1 billion gallons of bottled water in 2009. That breaks down to slightly under 28 gallons per person, according to the Beverage Marketing Corporation. That's down 3.2 percent from 2008, but that's still a lot of bottled water![13]

Bottled water, however, may not be any better than what comes out of your faucet. You may want to think twice before reaching for that pricey bottle of water as a safe alternative to tainted tap water.

First, bottled water isn't tested as often as tap water is for impurities and possible pollutants. Second, some bottled water basically originates as tap water, or comes from public reservoirs. PepsiCo. admits that's the case with its popular Aquafina brand. Coca-Cola Company's Dasani water also is from public water sources. A relative newcomer to the bottled world, Tap'd NY, advertises that it is New York City tap water—purified, of course.

WATER BATTLES

The bottled water business isn't without its water wars. One fight pitted the small town of Shapleigh, Maine, with about 2,300 residents, against Poland Spring Brand Natural Spring Water, part of giant Nestlé Waters North America.

Poland Spring wanted to draw waters from underground springs in the area, but in September 2008, town residents said "no," temporarily, and then made the "no" permanent in March 2009. Later that year, Poland Spring abandoned its test wells, and "moved on." No doubt there will be another fight another day.

A recent study by the nonprofit Washington, DC-based Environmental Working Group (EWG)[14] identified contaminants in some brands of bottled water. The group tested 10 U.S. bottled water brands in eight states and the District of Columbia, and turned up "thirty-eight different pollutants, including bacteria, fertilizer, Tylenol, and industrial chemicals," the group says. In releasing its October 15, 2008, report, EWG wrote:

> Laboratory tests conducted for EWG at one of the country's leading water-quality laboratories found thirty-eight contaminants. . . .
> The pollutants identified include common urban wastewater pollutants like caffeine and pharmaceuticals, an array of cancer-causing byproducts from municipal tap water chlorination, heavy metals and minerals including arsenic and radioactive isotopes, fertilizer residue, and a broad range of industrial chemicals. Four brands were also contaminated with bacteria.

As with most arguments over water, there are at least two sides to the situation. The International Bottled Water Association (IBWA), an Alexandria, Virginia-based industry group, points out what it says are flaws in the EWG report and disputes it in a press release:

> In the report, the EWG frequently mischaracterizes substances found in the tested bottled water products and discusses them out of context with accepted scientific determinations, [IBWA President Joe Doss said]. "In general, the report is based on the faulty premise that if any substance is present in a bottled water product, even if it does not exceed the established regulatory limit or no standard has been set, then it's a health concern."
>
> For example, EWG was critical of the bottled water brands found to contain fluoride. However, fluoride can prevent tooth decay and the American Dental Association has stated that, "Whether you drink fluoridated water from the tap or buy it in a bottle, you're doing the right thing for your oral health." Moreover, the levels of fluoride found in the bottled water tested by the EWG were all in compliance with the applicable FDA standards.[15]

No matter what side anyone takes in this bottle battle, it's clear that no water is safe these days from dispute. The rancor will only worsen as the water crisis grows.

WATER REALITIES

- The United States is one of the few nations in the world where 100 percent of its population has access to safe drinking water, yet thousands of Americans get sick from water every year.
- Even tap water isn't exactly "pure." An Associated Press investigation in 2008 found traces of pharmaceuticals like sex hormones, antibiotics, anticonvulsants, and mood stabilizers in the drinking water supplies of at least 46 million Americans.
- Lead, which is an unwelcome addition to drinking water, is present courtesy of antiquated infrastructure (pipes).

- Agricultural irrigation runoff carries with it fertilizers and chemical residues that create problems similar to those caused by urban storm water runoff.
- Man-made pollutants like the now-outlawed gasoline additive MTBE have found their way into drinking water supplies in some areas along the East Coast, including parts of New Hampshire.
- Nature provides its share of pollutants, too. Naturally occurring arsenic is a problem in areas of New Hampshire, New Jersey, Maine, and even California.
- Bottled water has its own issues, including contaminants in the bottle, according to studies.

NOTES

1. U.S. Environmental Protection Agency, Federal Register Environmental Documents, "National Primary Drinking Water Regulations: Ground Water Rule, 71 Fed. Reg. 65573-65660 (November 8, 2006), http://epa.gov/EPA-WATER/2006/November/Day-08/w8763.htm.

2. Michael F. Craun, Gunther F. Craun, Rebecca L. Cauldron, and Michael J. Beach, "Waterborne outbreaks reported in the United States," *Journal of Water and Health* 4, Suppl. 2 (2006):19–30, http://epa.gov/nheerl/articles/2006/waterborne_disease/waterborne_outbreaks.pdf.

3. U.S. Environmental Protection Agency, "Source Water Protections: Frequent Questions," http://cfpub.epa.gov/safewater/sourcewater/sourcewater.cfm?action=FAQ.

4. U.S. Geological Survey, "Man-Made Organic Compounds in Source Water of Nine Community Water Systems that Withdraw from Streams, 2002–05," http://pubs.usgs.gov/fs/2008/3094/pdf/fs2008-3094.pdf.

5. U.S. Geological Survey News Release, "Man-Made Chemicals Found in Drinking Water at Low Levels," December 5, 2008, http://water.usgs.gov/nawqa/swqa/GuidanceDocuments/pressrelease.pdf.

6. The cities included Albuquerque, New Mexico; Atlanta, Georgia; Baltimore, Maryland; Boston, Massachusetts; Chicago, Illinois; Denver, Colorado; Detroit, Michigan; Fresno, California; Houston, Texas; Los Angeles, California; Manchester, New Hampshire; New Orleans, Louisiana; Newark, New Jersey; Philadelphia, Pennsylvania; Phoenix, Arizona; San Diego, California; San Francisco, California; Seattle, Washington; and Washington, DC, Natural Resources Defense Council fact sheet, http://www.nrdc.org/water/drinking/uscities/popup_drinkingwater.htm.

7. Safe Drinking Water Program, Water Quality Control Division, State of Colorado, "Waterborne Salmonella Outbreak in Alamosa, Colorado ... Outbreak Investigation, Response, and Investigation" (November 2009), www.cdphe.state.co.us/wq/drinkingwater/pdf/AlamosaInvestRpt.pdf.

8. U.S. Geological Survey News Release, "Gasoline Additive MTBE Widespread in New Hampshire's Ground Water," January 2, 2008, http://nh.water.usgs.gov/WhatsNew/newsreleases/mtbe010208.htm.

9. U.S. Geological Survey News Release, "USGS Report Shows Where Arsenic Is Most Likely in New England's Ground Water," May 25, 2006, http://nh.water.usgs.gov/WhatsNew/newsreleases/arsenic052506.htm.

10. Agency for Toxic Substances and Disease Registry, "Camp Lejeune, North Carolina: Water Modeling," http://www.atsdr.cdc.gov/SITES/LEJEUNE/watermodeling.html.

11. Michigan Environmental Science and Services Division, "Preventing Groundwater Contamination," http://michigan.gov/documents/deq/deq-ead-tas-grwtrcon_329692_7.pdf.

12. U.S. Geological Survey, Toxic Substances Hydrology Program, "Land-Use Activities Release Naturally Occurring Arsenic," http://toxics.usgs.gov/highlights/arsenic_desorption.html.

13. Beverage Marketing Association, press release, "Bottled Water Confronts Persistent Challenges ..." (July 2010), http://beveragemarketing.com/redesign/includes/home/pressreleases.html.

14. Olga Naidenko, Nneka Leiba, Renee Sharp, and Jane Houlihan, "Bottled water contains disinfection byproducts, fertilizer residue, and pain medication," Environmental Working Group report (October 2008), http://ewg.org/reports/bottledwater.

15. International Bottled Water Association, press release, October 14, 2008, http://www.bottledwater.org/public/2008_releases/environmental.htm.

CHAPTER 5

ON GOVERNING WATER

There is water law and then there is water politics, and you can't separate the two.
> —Roger W. Sims, partner, Holland & Knight, specializing in water resources and environmental and land-use law

In the United States, water is allocated based on a complex system of regulations that vary drastically from place to place. Those rules are interpreted differently by different parties. Interested parties regularly fight over who gets what, when they do or don't get it, how much they get, for how long they will get it, and who is left out.

For the vast majority of Americans, turning on the faucet yields bountiful water even in areas suffering from drought or water shortages. Behind the scenes, however, it's very different, especially as demands for the resource grow and supplies dwindle. Water and the laws that regulate it are complicated at best and confounding at worst.

Water laws—some dating back to the colonial era—vary by locale, region, state, and beyond, and often depend on legal interpretations. Nonetheless, these laws are the legacy that water users across the nation have had to learn to live with in wet times and dry. Change is slow and, with so many stakeholders, tough to come by.

"The politics of water is one of the most intense and divisive issues I see in my practice," says Roger W. Sims, a water resources specialist in Orlando, Florida. As an attorney, Sims regularly deals with water management districts, state and federal environmental protection agencies, the U.S. Army Corps of Engineers, and the U.S. Fish and Wildlife Service. "It's largely because water is so essential, and people are used to having it fairly cheaply."

THE RIGHT TO WATER

Let's begin with a few basics of water law, keeping in mind that even the basics don't always apply to all cases in all places. As with oil and natural gas, just because you own the land doesn't mean you get the water. The water may be beneath your property, running through it, or bordering it, but that doesn't mean you get it! In fact, if you live in parts of the western United States, that water very well could belong to a person, business, or organization miles away. If such is the case and the water isn't yours, don't touch it or you could get in plenty of hot water, figuratively speaking. That's because of *water rights.*

Two Legal Approaches

Different parts of the country take different approaches to water rights. Water rights can be either *riparian* or *prior appropriation,* or a combination of both and more—or less. Prior appropriation, in use across much, but not all, of the West, is based on the concept of *first in time, first in right.* The first person or group to take a quantity of water and put it to full *beneficial use* has a higher priority of right than a subsequent user, according to the U.S. Geological Survey (USGS). In other words, the earlier in time that the water right was granted, the older, and more senior the right to access that water. An 1888 right granted and put to beneficial use until today takes precedent over a 1988 right, which in turn is senior to a 2008 right. As expected, though, the rule is not that simple, with plenty of nuances open to dispute. Beneficial use, for example, is a gray area.

Rights can also be lost if they're not used. Full rights or a portion of water rights may be sold, leased, or transferred separately from ownership of the land. In fact, the buying and selling of water and water rights (with plenty of restrictions) is a booming business in the western United States. We'll talk more about that in Chapter 7.

First in time is limited by the equally important requirement that a water user is only entitled to divert the amount of water that can be of beneficial use, says Clive Strong, a proponent of prior appropriation as Idaho's deputy attorney general and chief of its Natural Resources Division. The Idaho Supreme Court recently reaffirmed this principle. People today, especially critics, tend to forget the original purposes underlying the prior appropriation doctrine, says Strong. "First, the doctrine sought to encourage the development of the West. To do that, it limits the amount of water that can be diverted to an amount reasonably necessary to achieve the purpose for the diversion. This encourages optimum use of the limited supply. At the same time, the doctrine recognized that investment would be hindered without a secure water supply—thus the principle 'first in time, first in right.' Inherent in the doctrine is the tension between security of the right to the water versus the optimum use of that water."

Prior appropriation today makes more sense in states where water is in short supply, adds David R. Tuthill Jr., a longtime water expert and private Idaho-based water consultant with Idaho Water Engineering, LLC. "Anywhere there is scarcity, it provides us a mechanism for distributing water fairly. In states without prior appropriation doctrine, it's very difficult to administer available water supplies in a dry period."

The other approach to water rights is known as *riparian,* which originated from English common law. Historically much (but not all) of the eastern United States bases its water rights and law on this approach, which recognizes the right of use of a landowner whose property abuts the water. Specifically, if someone owns land next to a stream, riparian water rights give him or her right to reasonable use of that stream. A holder can't sell that right and must share the amount of water in that stream with others holding the same riparian right. Reasonable use is subject to debate. Riparian rights can differ from state to state, too, and have evolved in many places. See Table 5.1 for a state-by-state breakdown of surface water allocation approaches.

TABLE 5.1   States According to Their Basic Approach to Allocating Surface Water

Appropriate Rights	Dual Systems (appropriate and riparian rights)	Regulated Riparianism	Riparian Rights
Arizona[a]	Alaska	Connecticut	Alabama[b]
Colorado	California	Delaware	Arkansas[b]
Idaho	Kansas	Florida	Illinois[a]
Montana	Nebraska	Georgia	Indiana
Nevada	North Dakota	Hawaii[c]	Louisiana[d]
New Mexico	Oklahoma	Iowa	Maine
Utah	Oregon	Kentucky	Missouri
Wyoming	South Dakota	Maryland	New Hampshire
	Texas	Massachusetts	Ohio
	Washington	Michigan[e]	Pennsylvania
		Minnesota	Rhode Island
		Mississippi	South Carolina[a]
		New Jersey	Tennessee
		New York[e]	Vermont
		North Carolina[b]	Virginia[b]
		Wisconsin	West Virginia

Source: Joseph W. Dellapenna, professor of law, Villanova University: rapporteur, Water Resources Law Committee of the International Law Association: and director, Model Water Code Project of the American Society of Civil Engineers.
[a]The state has a regulated riparian system for groundwater, but not for surface water.
[b]The state has enacted a regulated riparian system but has largely not implemented it.
[c]Hawaii has a compound of ancient customary and prescriptive rights and regulated riparian rights.
[d]Louisiana follows riparian principles, but as derived from French law and expressed in its civil code rather than as part of the common law tradition.
[e]The state has enacted a regulated riparian statute that applies only to very large users on certain limited water sources.

Equal Rights to Water? As a whole, however, riparian water disbursement in the eastern United States entails a doctrine of public trust, according to Sims. "The water is held in trust for the benefit of the public, and the allocation of specific quantities of water becomes a governmental decision. The government—state, regional, water basin, or other water-controlling entity—says to one party, 'You can have this much water,' and to someone else, 'You can't have what you want.' There's really almost no federal interaction directly with the water's allocation under this system. In the West, on the other hand, you have the Bureau of Reclamation and Department of Interior (federal agencies) heavily involved in water disbursement."

Riparian water rights are intended to give people whose land borders on streams and rivers equal rights to the water, adds Chicago-based water issues expert A. Dan Tarlock. "However, nobody knows what the rights actually are because there has never been any real attempt to quantify them. Basically it's 'act first and see if you get sued,'" says Tarlock, a frequent consultant to government agencies, author of the treatise *Law of Water Rights and Resources*, and co-author of a casebook on water law, now in its sixth edition. "Every time I have to write an opinion for some water project in the East as opposed to the West, what I say is, 'There is no certainty here, so the real question is: What's the risk of getting sued?' and it's usually pretty minimal," Tarlock adds.

Regulated riparianism. There is no such thing as pure riparianism anymore, says Villanova professor of law Joseph Dellapenna. Instead, he offers a third approach to water rights, which he calls *regulated riparianism*. Dellapenna describes regulated riparianism as a system of public property in which the state manages the resource in trust for the public through time-limited permits. The permits provide the necessary water security for private action, while the time limits allow the state to revisit its decisions periodically—much like the way the Federal Communications Commission issues broadcast licenses or the Federal Energy Regulatory Commission issues dam permits.

This is, like traditional riparian rights, also a system of common property, but in that case each person with lawful access basically decides for him- or herself when, where, how, and how much water to use, says Dellapenna. Prior appropriation, on the other hand, is a system of private property in which each appropriator holds a water right that is strictly defined as to when, where, how, how much, and relative priority. Within the confines of these rights, appropriators are free to use the water according to their own judgment.

Some Eastern states have tempered their riparian rights approach with a water permitting system. Florida, for example, uses a permit system that works somewhat as prior appropriation does in the West, except that Florida permits initially are limited to a certain time and to "significant" withdrawals, adds Tarlock. As is the way with water, *significant* is defined differently in different places.

This approach does not involve prior appropriation, says Sims. "It's recognition of the rights of existing users to have a permit to continue using the water. At renewal time, however, permitted quantities are reevaluated and may or may not be approved. There are quantity thresholds below which you can use a general permit, and there's an exemption for private, domestic use."

"We are moving toward regulated riparianism," agrees Philadelphia water expert and attorney Kenneth J. Warren. "Under the old riparian doctrine, courts applied a reasonable-use standard. With surface water, at least, a person owning land abutting the stream could use the water as long as the landowner did not unreasonably interfere with other users of that surface water. With the need for public water systems and economic development, that system evolved, and today many jurisdictions employ a permitting system," says Warren. "In the Delaware River Basin, for example, the Delaware River Basin Commission issues permits for large withdrawals. Permits are for a limited duration and, unlike prior appropriation in the West, the regulatory agency can modify or revoke them. That's the case even though some permits do not have an automatic expiration date and may look like they are permanent. It is the right of the agency to modify or revoke the permit."

Portent of water wars to come. Permitted water rights in the East are not granted in perpetuity, Sims adds. "Other users can come in and compete for that water. So far, at least in Florida, the water management districts have been able to resolve competing-use conflicts. But it's a war waiting to happen. No specific rule really explains how you sort out competing uses. The statute says [that] all other things being equal, the renewal application gets the preference."

How does a water management agency determine if all other things are equal? Like so many questions and issues relating to water, the solution is open to interpretation.

Another state that has adopted a water permit system for significant withdrawals is Massachusetts. The Massachusetts water management act requires the state environmental protection agency to consider whether a large new water withdrawal is a good idea from the standpoint of the public interest before it issues a permit. The agency must determine, for example, whether in-stream flows will be depleted and whether wildlife will be adversely affected when it considers permitting a new water project involving sufficiently large withdrawals, says Lee P. Breckenridge, a professor at Northeastern University School of Law in Boston and an expert on environmental and natural resources, land use, and property law.

DISPUTES AND DECISIONS

As expected, disputes and subsequent decisions over water rights vary between East and West—and among states that follow the riparian or prior appropriation doctrine, or some derivative of one or both. Architect and urban design sustainability expert Daniel Williams comes down on the side that opposes prior appropriation. "What we are doing right now is dangerously misinformed!" he says. "For the last few hundred years they've had water rights, they're holding everybody hostage."

Adjudications in the West

In the West, the decisions over water disputes are known as *adjudications* and, also as one might expect, the methods of adjudication vary by state. The controversies are unending, and the results of adjudications seldom satisfy both parties.

As mentioned earlier, in Arizona, for example, a group of pecan farmers lost their livelihood when their trees died of thirst after their aquifer level dropped by half from 32 feet to 16 feet. The farmers blamed the depleted aquifer on an industrial neighbor that drew down the aquifer. However, adjudication said their neighbor was entitled to all the water it took.

Similarly, more than 400 homeowners in a Colorado subdivision ended up without water when their wells were plugged after adjudication found that they had no rights to the water under their land.

Court Appeals in the East

Parties in states that take a riparian approach to water do not rely on a stream adjudication that divides the right to use water among all claimants based on the date each claimant first appropriated the water and put it to beneficial use. Instead, they generally pursue individual appeals in court if they believe they've been adversely affected by the granting or denial of a water-use permit, says Warren. "We don't take an entire stream flow and divide it among all claimants in a manner akin to property rights, as the West does in [its] stream adjudications."

It is also highly unlikely that an individual homeowner in the East who seeks to drill a well to supply a residence would be denied the right to withdraw water, adds Warren. The water regimes in the East generally assume the existence of sufficient supplies. The amount of water withdrawn by an individual homeowner in most instances is unlikely to interfere with needs of other users and may be below the quantity thresholds that require permits. "Thresholds," naturally, vary by water basin or regulatory agency, region, and more.

WESTERN WOES

Ownership of water in the West generally is separate from land ownership and is quantified in terms of entitlement to a certain number of gallons or acre-feet at a certain time or for a certain period. Just as someone buys a piece of land for the right to build or own a home on the surface of that land and acquires the surface rights, each piece of land across the country also has separate mineral rights, and, in the case of the West, separate water rights. Those rights to water can be further divided into surface water, groundwater, and various aquifers' water; divided again into primary, secondary, tertiary, and gray (wastewater) rights; and so on. A rigid pecking order determines who gets the water first, how much of it he or she gets, and so on down the line. Primary or senior water rights are first in line, followed by secondary, and so on. Of course, that's unless water treaties or agreements are in place that state something else.

The most contentious water battle Strong says he has seen in 25 years in Idaho pits water rights users against each other. The fight came to a head—or so people thought at the time—in March 2009 when senior surface-water rights holders claimed that groundwater pumpers with junior water rights should be curtailed (stopped from taking water).

The sides failed to reach a compromise, and Tuthill, then director of Idaho's Department of Water Resources, issued a notice of curtailment to the junior rights holders. That meant the junior rights holders would lose their primary water supply if they failed to provide an acceptable mitigation plan. The thirsty junior rights holders, instead, offered an alternative plan that included building a pipeline to provide the senior rights holders with the water they are entitled to under the state's water laws. Tuthill accepted the plan. A done deal? Not by any stretch of the imagination. Both sides agreed to delay the pipeline, but the junior rights holders failed to comply completely with other aspects of the mitigation plan. Says Tuthill, now retired from his 33-year public service career: "My expectations are the state Supreme Court will rule on all these rulings. These are fundamental issues for Idaho, and the law is not crystal clear."

These adjudications [decisions over water], specifically the senior rights holder (a trout farm) versus groundwater pumpers (farmers) with junior rights, represent what Tuthill refers to as conjunctive administration of water. It's all about dealing with the relationship between groundwater and surface water. "In the last century we've been successful administering surface water, but ignored the impact on groundwater of surface water pumping across the West," he adds. "Now that's changing with the result that senior surface-water rights holders will get some relief from the water-depleting impacts that have been occurring over the decades as a result of groundwater pumping."

Naturally, it's not that clear and simple either. A municipality like the city of Boise, Idaho, which primarily uses groundwater as its water source, may have junior rights to the water, *but* the domestic portion of a water right can condemn (take precedent over) other uses. Nondomestic uses like lawn irrigation, however, can't be condemned without other mitigation. Conjunctive administration will encourage water banks and other forms of water marketing and trading as groups scramble to find the water they've always taken for granted, adds Tuthill.

Resource Shortage

Water rights aside, there are still more variables. Access to the water also can change depending on current climate conditions, the amount of water available, and whether conditions are such that the water can even be distributed as allocated by the water rights. Climate change and drought figure into the equation because drought can affect the successful transportation of water to the owners of its primary rights. Remember Colorado rancher John Gandomcar, whom we talked about earlier? On one of his properties, he has junior rights to a set amount of water that runs in a nearby creek. However, Gandomcar's rights are secondary to those of the senior rights holder—a farmer miles down the creek. Therefore, if water is in short supply, Gandomcar is out of luck. Or is he? What happens if the Eastern Plains of Colorado—and the farmer's fields—are parched by drought, and the farmer with senior water rights "calls" the river? That means he demands, or *calls for* his water allocation to be delivered via the creek as dictated by his senior rights. But because the air and land—including the often-empty creek bed and surrounding ground—are so dry, the water won't be able to make it to its rightful owner, despite his senior rights. So Gandomcar's secondary rights win out because he's closer to the source and therefore can utilize the water as opposed to wasting it.

That's what's known as a *futile call*, says Tuthill. The senior rights holder calls on someone else's junior right, but curtailing or taking away the junior right does not result in any improvement in the water right for the senior. "It would be futile for the downstream senior right holder to shut off the upstream junior because the water wouldn't get there anyway," Tuthill adds.

Confused yet? It gets more complicated.

Overallocated

The overextended Colorado and Rio Grande river basins are prime examples of the perfect storm—the water battle waiting to explode—when it comes to water rights in the West. A total of 18 million acre-feet of annual Colorado River water flow has been allocated to various entities, according to Southern Nevada Water Authority's Pat Mulroy. Yet it is estimated that the river's annual flow today is closer to 11 million to 12 million acre-feet a year, especially when one factors in evaporation.

"There are more water rights, more legitimate claims of need and use, in these areas than can possibly be satisfied under present hydrological conditions," says Colorado and Rio Grande basin expert Steve Mumme. "Part of that was bad planning, a little of it was bad law, and most of it was just a poor understanding of Mother Nature. Increasing development has placed increasing demands on the Colorado and Rio Grande river basins. And, in the last fifty years, we have become much more attentive to the quality as well as the quantity of water. We can't just deliver 1.5 million acre-feet of saline water to Mexico. That's not legal under the interpretation of the 1944 water treaty."

Regional demand for water is increasing just as Mother Nature has stopped cooperating, Mumme adds. "Whether that's due to systemic human-induced climate change or other processes independent of human beings is immaterial. The fact is there is less water with much higher demand, so that means stiff competition for the resource in the West," Mumme says.

It's that world water equation again:

Growing demand + Limited supply = Shortage and conflict
over what's left

SHOULD YOU CARE?

The average person pays little attention to all this mumbo-jumbo about water rights. If water comes out of the faucet when you need it, that's all that seems to matter. That is, unless your lawn looks parched one day, so you turn on the outdoor faucet and end up with a ticket for breaking the law—in this case, watering your lawn in violation of restrictions imposed because of a water shortage. That's hardly a far-fetched scenario. In fact, it's reality in many places coast to coast in the twenty-first century. If you haven't yet faced voluntary or mandatory water restrictions where you live during at least part of the year, you may well face them in the future, experts agree. Beware the water police! If you water when you're not supposed to, they could fine you, at the very least.

What Are Your Rights?

If you live in the West and own your home, you may want to research your water rights, too. Mineral and water rights to land on which homes, schools, businesses, and more are often owned by literally dozens of other people. Some subdivisions even were (and are) being built on land that doesn't have full senior rights to steady water supplies. As more and more people move to arid portions of the country with limited water supplies, they risk ending up high and dry in the event of a water shortage. (You may not own that rain falling from the sky on your property, either!)

MORE VARIABLES

Water rights are further complicated by treaties, compacts, and water dispersion and diversion deals—and not only in the cases of the Colorado and Rio Grande rivers mentioned earlier. Don't forget Massachusetts' overallocated Ipswich River, which ceased flowing to the sea in the summer.

The Law of the River

Water-related laws may not make sense to some people in the context of today. Yet they are the law of the land—or the water— and demand adherence. Consider the Republican River that runs through Colorado, Nebraska, and Kansas. Ownership of the land— specifically, the river bottom—doesn't necessarily equate to ownership of the water in the river. Ownership of the water is divided based not on how many miles of river are in each state, but on a contract signed in 1943.

At that time, the state of Colorado sold off the right to 89 percent of the annual flows of water through the river— 49 percent to Nebraska and 40 percent to Kansas.

The state of Colorado then could face lawsuits and hefty fines if it takes more than its 11 percent share of the water in any given year. And that's the *law*—no matter how large Colorado's population grows, how thirsty Coloradans get, and how parched the land becomes.

Remember thirsty New Berlin, Wisconsin, near Lake Michigan, the town that's split by a subcontinental divide? Half that city's water drains into the Mississippi River Basin and the other portion drains into the Great Lakes Basin. As we mentioned, the problem New Berlin faces, and the reason it can't just suck water from Lake Michigan to quench its thirst, is that strict laws govern moving water from one drainage basin or watershed system to another, including international laws such as the recently enacted Great Lakes Water Resources Compact.

DAVID VERSUS GOLIATH

Laws and the courts aside, the classic David versus Goliath water battle—complete with dynamite, shotguns, soldiers, and armed rebellion—dates back to the early 1900s. It pits the mighty, thirsty Los Angeles represented by William Mulholland, chief engineer of the city's Bureau of Water Works and Supply, backed by the federal government, against the farmers of tiny, then water-rich Owens Valley about 250 miles north of Los Angeles. Guess who won? Here's a clue: It wasn't the little guy.

Mulholland, along with former Los Angeles mayor Fred Eaton, envisioned tapping the abundant water in Owens Valley for fast-growing Los Angeles by building an aqueduct to transport the water. It was a project marked by deceit, dishonesty, and violence. The aqueduct was blown up a number of times, and Owens Valley

WATER TALES

The Owens Valley saga isn't over yet. Organizations still offer updates and information on its history, changes, and restoration. To learn more, visit

- The Owens Valley Committee: http://ovcweb.org/
- Owens Valley History: http://www.owensvalleyhistory.com/

locals even shut off water to Los Angeles for a time. Nevertheless, Mulholland got his way, and Los Angeles got its water, at the expense of what became a dusty Owens Valley. "As water lawyers will tell you, water runs uphill to money," says Steve Mumme.

Showdown in Nevada

Fast-forward to today. One of the biggest contemporary David versus Goliath water wars pits Snake Valley, Utah, against the Southern Nevada Water Authority (Las Vegas). The battle has brewed for years and could continue to brew for years more in the legal arena under the jurisdiction of Nevada's state engineer. The Southern Nevada Water Authority requested 52,000 acre-feet of water—about 17 billion gallons of water a year—from the Snake Valley, which straddles Nevada and Utah. Tracy Taylor, then Nevada state engineer, was set to hear arguments in the case in fall 2009, but that was delayed until fall 2011. The two sides sat down to hammer out a water-sharing agreement outside of the courts. Then the Nevada Supreme Court in 2010 tossed out some of SNWA's claims to water rights. Meanwhile, Nevada's new state engineer, Jason King, has laid out a timeline to reconsider the rights. There's even a web site devoted to developments in the fray (http://water.nv.gov/hearings/water-hearing/SpringDryCaveDelamar/index.html).

The battle for this "clear gold" amid an arid, thirsty West is contentious to say the least. "By far the greater amount of current use of the water is in Utah, while most recharge occurs in Nevada," says Hugh Hurlow, senior geologist for the Utah Geological Survey and a scientific adviser to the negotiations between the two parties. Because the applications to appropriate the water are in Nevada, it's the Nevada water engineer's decision even though the water runs off to supply the Utah side of Snake Valley.

"There is a certain amount of developable groundwater that could be taken out of the basin without a disastrous impact on the local economy," says Hurlow. "However, I doubt that it is as much as southern Nevada would really want out of the project."

As a scientist, Hurlow has spent the last few years studying the area's groundwater and geology, and how the proposed withdrawal might affect the groundwater flow. "Technically, the withdrawal would be legal, but there are consequences for Utah," he adds. "The greatest impacts of large-scale groundwater withdrawal and consequent lower water levels in the Snake Valley would be [on] irrigation operations and reduced discharge of springs that provide habitats for sensitive species and grazing rights."

Owens Valley revisited? "In one sense we have a David versus Goliath," says Jerald Anderson, a longtime farmer on the Utah side of Snake Valley. He also holds degrees in physics and accounting, and is a 25-year veteran in computer systems and construction management. Anderson recounts what he's seen water-wise since joining his dad on the ranch about a dozen years ago.

> When I first came out to work this farm in 1996 [Anderson's father had bought it 10 years earlier], we were flood irrigating. My dad was a civil engineer, and he had planned all this—done soil tests, water analysis, and devised a system that worked very well for irrigating the ground.
>
> A year later, we started into a period of drying. It wasn't drought yet, but we started to have less recharge, less rainfall. What had happened weather-wise in the 1980s was a very wet period that included flooding and the creation of new lakes, all of which recharged the aquifer. Over the next years, we used water that had accumulated because of that particular wet period.
>
> But starting in 1999, the irrigation system we were running couldn't keep up with the dryness of the land anymore, so we started being unable to flood irrigate. In 2003, we put in a pivot irrigation system. The drying forced us to change our irrigation practices, not just because there wasn't the water to pump, but because the ground was so dry and the climate had dried so much that we simply couldn't move the water the way we had before.

"It's not that we don't want Las Vegas to have water," Anderson says. "It's that we sincerely believe that science and experience will prove that if they try to take this water, it won't work. We would rather they understood everything before they messed it all up and then said, 'Oh, well, we'll have to fix that later,' like they did in Owens Valley in California back in the 1920s.

"Although the Southern Nevada Water Authority assures Utahans that won't happen here, we live on this land; we use the water," Anderson says. "We see how it responds to the recharge and the weather, and we would prefer it not be turned into a dust bowl by reducing water levels to the point where the vegetation can't survive out here," he says.

Anderson doesn't believe there is a sustainable resource at the level Las Vegas is seeking. "We're going from one of the wettest periods in history to drying out," he says. "Every chart I have of our water levels shows us losing groundwater every year. In the long term—75 years—this water isn't going to be here for anyone's use. They will have taken more than can be sustained, and they'll be without the resource."

The Snake Valley is a shared basin, and "how much of that water will be available for Utah use and how much for Nevada use is really the crux of the issue," says Southern Nevada Water Authority's Mulroy.

"What has historically happened in the West—and the most poignant example of that is along the California border—is that water was simply available for the taking. As a result, California has dewatered many of the shared basins between Nevada and California. So I have no problem with the notion that we have an obligation to environmental stewardship and that many issues must be addressed in Snake Valley," she adds.

As an example of those issues, Mulroy points to Spring Valley, where the Southern Nevada Water Authority owns groundwater rights. It has filed for unused perennial groundwater in that valley and, to protect the environment, has agreed to co-manage the basin with the U.S. Fish and Wildlife Service, the Bureau of Land Management, the National Park Service, and the Nature Conservancy.

"The only reason we bought that Spring Valley water was not to export it but to take that surface water as it comes in the spring melt and artificially recharge the groundwater basin there," says Mulroy. "We bought ranches close to springs. . . . Ranching isn't our objective. Our objective is to use those ranching water resources to manage the ecosystems."

There's a reason that *groundwater* is a single word, adds Anderson. "You can't separate the ground and the water, and I think that's an important concept. This part of the West is the way it is because the water is where it is. When you start moving the water around, then something has to change."

Anderson says a major difference between Utah and Nevada is that decades ago Nevada allowed people to take the water, even at the expense of the desert drying up. Utah, on the other hand, emphasized that water must be available if the environment is to flourish. "So there is a philosophical difference between the two states in the way the water is administered," Anderson says.

Adds Mulroy, "We have filed applications for unused, untapped perennial yield groundwater that will afford us the opportunity—with continued conservation—to keep this city alive in the event we lose Lake Mead [which provides about 90 percent of Las Vegas' water]."

Whether Las Vegas wins the battle with the "north" for the water depends on many variables. The U.S. Supreme Court likely will end up considering this case, because it's an interstate issue—the Snake Valley lies partly in Nevada and partly in Utah. Adding to the fray, Snake Valley water generally is replenished in Nevada's mountains, but the water flows north into Utah.

In the meantime, Mulroy and her agency continue to "diversify" the origins of their water.

For more information from various parties involved in Las Vegas' pursuit of water, check out a few of these sources.

- Great Basin Water Network: http://greatbasinwaternetwork
  .org
- Millard County Chronicle Progress: http://millardccp.com/
- Protect Snake Valley: http://protectsnakevalley.com
- Southern Nevada Water Authority: http://www.snwa.com

LAWS CREATE STRANGE BEDFELLOWS

The use-it-or-lose-it approach to water rights in the West makes for unusual business partners. The Southern Nevada Water Authority, for example, has gone into the cattle-ranching business. But not because the water agency is strapped for cash, looking to diversify its holdings, or has a burning desire to get back to the land. It's in part Western water law's dictum of "use it or lose it." The district purchased 23,000 acres of ranch land in northern and central Nevada, which came with hefty water rights. However, if the district doesn't use those rights, they're history.

The Southern Nevada Water Authority isn't the only ranch-owning utility in the Southwest. El Paso (Texas) Water Utilities claims close to 75,000 acres, divided among two ranches and three big farms, on its balance sheet, according to Ed Archuleta, the utility's executive director. "Like many cities, we have invested in farms and ranches outside El Paso County, so we have the water rights and could import water in some crisis," says Archuleta. "That's an insurance policy for the future. We acquired the land now because in Texas, you must own the land to have the right of capture underneath for the groundwater."

FEDERALLY RESERVED WATER RIGHTS AND THE NATIVE AMERICAN EQUATION

One big—and definitely contentious—variable in Western water allocation laws is and has been federally reserved water rights. They are associated with most lands managed by the federal government, including national parks and monuments, national forests, military reservations, and Indian reservations. According to the U.S. Bureau of Land Management (BLM), and affirmed by the Supreme Court, these lands are mostly immune from state water laws and thus are not subject to the typical rules dealing with diversion, beneficial use, and nonuse requirements.

By far the biggest unknown when it comes to federally reserved water rights is the Indian reservations in the West. The tribes on these reservations are a source of real "water power"—and confusion. When the federal government established the reservations by "reserving" the land and the right to it on behalf of the tribes, it also reserved rights to sufficient water. The meaning of *sufficient* has been open to discussion.

"Native American water rights have been pushed under the rug," says Sandia National Laboratories' Mike Hightower. "They're not discussed openly, but they could be a major bone of contention in the future."

In the *Winters* Doctrine, stemming from a 1908 Supreme Court case, *Winters v. United States*, the federal government reaffirmed this federally reserved Native American right to current and future quantities of water for their reservations. The crux of the matter is that Indian water rights in many cases are senior to prior appropriation rights because their date of priority is the date the reservation was established. However, it's not that simple. Among the confusions, the Supreme Court in *Winters* found that an Indian reservation may reserve water for future use in an amount necessary to fulfill the purpose of the reservation. The amount of water, then, is left open to interpretation, challenge, and more confusion. That also left the federal government smack in the middle of states' water management. To remedy that somewhat, in 1952 Congress passed the McCarran Amendment, which stipulated that these and other federally reserved water rights are subject to state adjudication. In the past several decades, Indian tribes and state governments have worked to negotiate settlements on the quantities of water they're *each* entitled to, but many cases remain unsettled.[1]

If you're wondering what kind of precedence and "water power" American Indians can wield, you had better believe they're substantial. In the case of the Wind River Indian Reservation and reserved rights to water in the Big Horn River system in Wyoming, the Big Horn General Adjudication awarded the Shoshone and Arapaho tribes of the reservation the right to 500,000 acre-feet of surface water annually with a priority date of 1868![2] That's a major and very senior water right—and that means it trumps all rights after it. To put that amount of water in perspective, one acre-foot equals 325,851 gallons of water. That's nearly 450 million gallons of water a day. The city of Albuquerque, New Mexico, with a population of about 500,000, consumes about 50 million gallons of water a day, Hightower notes.

Potential Native American reserved rights could create a nightmare scenario for another river—the Colorado. Under the *Winters* doctrine, says Hightower, Indian tribes with large land holdings claim the right to significant water in the Colorado. That's a colossal problem because there may not be enough water in the Colorado River to meet the actual demands of the *Winters* doctrine, he adds.

PUBLIC OWNERSHIP

Still other variables affect water rights. Some water, even in the West, is considered to be in the public domain. The public has the right to the water or at least to some of it in some capacity. That's the case with some, though not all, water on public lands. However, there's a catch-22 in all this, too.

In California, for example, and elsewhere as determined by the U.S. Supreme Court, the public has access rights to navigable waters no matter who owns the rights to drink or use the water. That can have both positive and negative effects on water supplies and access to them for drinking and industry. The California Supreme Court used that principle to limit the amount of water the city of Los Angeles can divert as drinking water from certain tributaries of Mono Lake in the northeastern portion of the Sierra Madre Mountains. Because the rivers are used for recreation, a certain amount of flow is required to remain in the river, thus limiting what Los Angeles can draw out.

THE U.S. GOVERNMENT'S ROLE

The federal government also joins the party at the water-distribution table because of its responsibilities for broad regulation and monitoring of the resource. Think Hoover Dam in the West, the Clean Water Act, and more.

It's tough to get a handle on the water issue when more than 20 different federal agencies have a hand in water use, development, and management, says Sandia Laboratories' Hightower. He points to a few agencies and some of their tasks.

- USGS handles groundwater.
- The Bureau of Reclamation regulates surface water and hydropower in the West.
- The Defense Department's Army Corps of Engineers manages flood control and navigation.
- The National Oceanic and Atmospheric Administration (NOAA), part of the Department of Commerce, deals with precipitation, rainfall monitoring systems, and more.
- The Department of Agriculture regulates irrigation and rural water and electric utilities.
- The National Nuclear Security Administration's national laboratories study water as it relates to national security issues.

- The Environmental Protection Agency (EPA) regulates water quality.
- The Tennessee Valley Authority and the Bonneville Power Authority regulate hydroelectric power.
- The Fish and Wildlife Service oversees water issues related to endangered species and ecology.

This list includes federal government agencies only, says Hightower, and doesn't include the state, regional, and local entities that manage water in their specific areas. The lack of coordinated efforts among all these agencies definitely creates problems and hampers coordination, he adds.

The pot with the greatest number of hands in it just might be in the Colorado River Basin. In addition to federal agencies—Lake Mead is a project of the Bureau of Reclamation—the basin covers the seven states (Arizona, California, Colorado, Nevada, Utah, New Mexico, and Wyoming) that signed the Colorado River Compact of 1922 (http://www.usbr.gov/lc/region/g1000/pdfiles/crcompct.pdf). That compact "provides for the equitable division and apportionment of the use of the waters of the Colorado River System." Known as the "Law of the River," the agreement also involves Mexico and a host of cities, water districts, and agriculture entities on both sides of the border.

LAWS: PROBLEM OR SOLUTION?

Many water experts agree that in some instances, laws related to water may actually exacerbate issues and problems in the United States.

Existing water laws, especially regarding prior appropriation in the West, have made addressing many environmental issues very difficult, says Tarlock. That's because resolving certain environmental concerns requires that additional water remain in streams. But, Tarlock adds, "because the water is already spoken for—allocated based on water rights—that would potentially mean taking someone's property and saying that they can't use the water they own the rights to because it must remain in the stream."

Another criticism of prior appropriation is that because it's a "use it or lose it" approach—if you don't use your full allocation under your water rights, those rights can be revoked, forfeited, or abandoned—people overuse rather than conserve water, Tarlock says.

Today more than ever, the management of water in the United States and Mexico is all about water rights, and that to some extent is part of the problem, says Steve Mumme. He adds, "There is a certain institutional complexity here that really affects the little guy. The way water is used and the way it is consumed are directly dependent on a welter of overlaying U.S. federal, state, and even local laws that are complicated by international treaties and agreements. . . . We have created systems in which, especially in the Southwest and the more arid regions of the country, the use of any land or real property is tied to water property rights. Any changes in these laws, and especially changes to the treaties, affect these property rights. . . . If you start mucking around with the foundational documents, everybody is affected, and everybody goes ballistic."

LEGAL OVERHAUL AHEAD?

An overhaul of existing water laws obviously isn't a popular or, for that matter, an especially practical or immediate solution. "All things considered, don't expect radical changes in water rights and laws. We have so many stakeholders with vested interests that it will be difficult to make dramatic changes unless they are confronted with a crisis," says Frank Richards, a former hydrometeorologist with the National Weather Service's Hydrologic Information Center.

However, change is absolutely necessary, says Villanova's Dellapenna, author of *Water and Water Rights*, a standard reference work on U.S. water law. Revision is necessary, because existing water laws are premised on an expectation about the relationship between a certain amount of water supply and demand. "That expectation no longer holds," says Dellapenna. "This is most dramatic in places like the Colorado River Basin, though it's true across the country to varying degrees.

"First, we need water law reform at the state level," says Dellapenna. "Second, we need a better system of state cooperation and coordination. States, for the most part, have not had a very good history in this regard."

The Political Arena

It's also tough to initiate any radical change, because of various prior commitments, adds Mumme. "You can't find a much more politicized arena than water and water rights. It's really at the foundation of wealth and the creation of wealth and the subsistence and well-being of society. So you have to acknowledge its complexity, and you end up tinkering with it; you make minor adjustments."

Those "minor adjustments" manifest themselves in all kinds of creative and not-so-creative ways. "There is room in the laws for increased efficiency requirements," says Mumme. "In theory you can only use water *beneficially,* which means not wastefully. We've been pretty lax in defining what waste is. You could boost the standards for what is waste, for example."

Philadelphia water law expert Kenneth J. Warren agrees. "*Beneficial use* is largely a water quantity doctrine, and the EPA has relatively little authority over water quantity, so you have to ask, 'What are the states doing?' I do think [that] in that context, there is substantial opportunity for states to better define *beneficial use,* and they have an incentive to do so in locations where water is scarce."

Federal, Regional, or Something Else

Some areas of water management can be solved only at the federal level, adds Warren. "I don't think you're likely to see the Colorado River, for example, regulated on a state-by-state basis. So the question then becomes, 'Do states form regional agencies to regulate interstate flows as we did in the Delaware River Basin with the Delaware River Basin Commission, or does the federal government intervene?' My personal preference would be to have regional organizations regulate. Having the regional flexibility to adopt solutions suited to the affected communities is better than having a one-size-fits-all national program. But I do not think you can leave it to individual states in all instances. Some of the problems are just too great for any one state to handle on its own."

"I expect every state would fight valiantly to not lose the authority to regulate its water," adds Tuthill. "It's an authority that's been entrusted to the states, and nationalization would meet huge resistance. Don't forget that national support for infrastructure played a big role in development of the West. . . . Structures built by the Bureau of Reclamation and Army Corps of Engineers [think Hoover Dam] have been very important in the development of the West and they continue to be important in the economic climate and fabric of the West. The day of building important structures should not be ended, either."

Water quantity control. However, there is no national plan for the federal government to take over regulation of water quantity from the states, says Warren. "State law largely prevails, and I know of no federal initiative to override state law. But EPA has acknowledged the advantages of what it calls watershed management, which is a significant advance from more traditional end-of-the-pipe or piecemeal analysis," he says. He likes a holistic approach to watershed management. Historically, agencies like the EPA regulate water quality but have virtually no regulatory authority over water quantity. States have traditionally handled water quantity in part by statute and in part through the common law.

"The modern view is that water quantity and quality are intimately related to each other," says Warren. "So to have two separate regulatory regimes to govern quantity and quality doesn't work very well."

For that reason, regional agencies like the Delaware River Basin Commission, which has authority to regulate both water quality and quantity, present a good structure for integrated management of regional water resources, Warren adds.

The Delaware River Basin Commission and the Susquehanna River Basin Commission are two of the best examples of interstate cooperation over water issues, concurs Dellapenna. Both have been established by compacts that spell out their powers. "They can regulate what states do; they can regulate what the federal government does; they can regulate what individual users do. That's a model for others," says Dellapenna. "The commission is composed of one representative of each state plus a representative of the federal government—each with one vote—so the federal government can be outvoted. Both commissions strive for consensus. Given that emphasis, they don't necessarily reach the best possible solution. But at least they act jointly; they aren't at cross-purposes; they aren't injuring each other. It's a definite step forward."

The Great Lakes Compact. One of the more recent much-discussed interstate water agreements is the Great Lakes Basin compact that former President George W. Bush signed into law in 2008. "It's a good idea for the present," says Tarlock, who worked on the compact.

Dellapenna, however, questions its efficacy. Participants in the compact agreed to set basin-wide standards collectively. But they left enforcement of those standards up to individual states. "Let's hope it works. Obviously that is potentially a huge, huge problem if one state turns out not to be enforcing the standards the way other states think it should," says Dellapenna. "In theory they have a council that can override decisions, but I think that will be a hard sell. . . . These commissioners very clearly represent individual state governments.

"So what is going to happen in the Great Lakes Council?" he asks. "Perhaps in a very egregious case the council might act, but for the most part they are likely to say, 'If I jump on Illinois for doing that, then they'll jump on me for doing something else.' It's a can of worms."

It gets tougher. Whatever happens with the Great Lakes and beyond, it's clear that decisions on water supply issues are likely to become more difficult as supplies dry up. The solutions are likely to become more and more expensive, too, says Orlando attorney Sims. "Water is such a controversial subject that any change in the laws is going to vary state to state, depending on the politics of the state. Those areas that have water really don't want anyone managing their water out from under them. And those that don't have it obviously would like some state or federal help to make water from water-rich areas available to them. . . . Given enough time and enough urgency, though, there may be enough political wherewithal to make something happen at a centralized level, but I think it would be extraordinarily controversial."

When the crisis is big enough, people will be more open to changes in water laws, adds Dellapenna. "If there is a severe enough crisis, most people come to their senses. We have seen that in lots of Eastern states where they have abandoned traditional riparian rights. In normal times when there is not a severe crisis and some-body talks about water law reform, all existing water users get all huffy and say, 'You can't take away my property.' But in a severe enough crisis, everyone agrees that we have to do something."

WATER REALITIES

- The two basic approaches to water rights are the traditional riparian right, stemming from English common law and in effect in much of the East; and prior appropriation, which conveys private ownership of a right to use water based on first in time, first in right.
- A third approach is a combination of both, which is sometimes called *regulated riparianism.*
- In the *Winters* Doctrine, the federal government reaffirmed American Indians' rights to current and future potential quantities of water for their reservations. Complicating the picture, however, is the fact that the government didn't spell out how much water Native Americans were entitled to.
- The federal government is involved in water quantity, quality, and control with its Bureau of Reclamation and Army Corps of Engineers projects as well as through more than two dozen agencies that in some way regulate or address water issues.
- Many water experts agree that in some instances, laws related to water may actually exacerbate issues and problems in the United States.
- Change isn't easy, especially because existing laws and processes are firmly entrenched and often involve financial gain, and obtaining consensus is tough.

NOTES

1. U.S. Bureau of Land Management, "Western States Water Laws: Federal Reserved Water Rights," http://www.blm.gov/nstc/WaterLaws/fedreservedwater.html.
2. Wyoming Water Development Office, Wyoming State Water Plan, "Wind/ Bighorn River Basin Plan, Executive Summary," http://waterplan.state.wy.us/ plan/bighorn/execsumm.html.

AMERICA'S WATER GODS

When it comes to the water gods—the courts—are the mitigators.
—Bruce Newcomb, former speaker of the
Idaho Legislature (1998–2006) and
longtime Idaho rancher

Bow to those who control the water, for they decide who gets it and who doesn't. Unfortunately, that's the current reality. Those who control the water, and their actions (or inaction), decide the fate of water across the country. Outside of the courts, they often interpret how the laws are carried out, and they execute the adjudications, court rulings, laws, and numbers. They know how to work within the existing water laws that determine whether a developer gets water to build his community, a farmer is entitled to enough water to sate his livestock or fields, or a city gets the water its residents need to live an unrestricted life water-wise.

Those with water power can be groups or individuals: some head water-governing agencies, some are water engineers, some are skilled private citizens, perhaps politicians or public figures, and others just maneuver behind the scenes. Some do so for the good of the community and Earth's resources; others have some other agenda—private or otherwise. These are the water elite.

WATER POWER

State engineers like Jason King in Nevada, who decides whether Las Vegas gets its water; Pat Mulroy in Las Vegas, whose quest for more water for her city is legendary; and Ed Archuleta in El Paso, Texas, a leader in making water conservation work, are all examples of the water elite. Plenty of other state engineers and directors of resource departments, along with heads of municipal utilities and watershed and water-management districts, also qualify for membership in the water power elite.

Ted Turner may be another water power to watch. Environmentalist, rancher, and the founder of CNN (Cable News Network), Turner also is the largest private landowner in the country, with about 2 million acres in holdings—and holds massive water rights associated with his land. Turner is a visionary (consider his quest decades ago to create a 24-hour cable news channel). Though he's made light of it in interviews, talk abounds that he's buying land in part for water rights and eventually will have significant water holdings. In February 2008, he told a reporter at the *Omaha World-Herald* that he didn't intend to sell any water rights to anyone. Time, of course, will tell.

Public water suppliers can also be part of the water elite. In Massachusetts, for example, they have a lot of power and exercise undue influence, says Kerry Mackin, who heads up the comparatively small Ipswich River Watershed Association that we talked about in Chapter 2. Mackin points out that the Massachusetts Department of Environmental Protection (MassDEP), the state's regulatory agency, has acknowledged that it treats water suppliers' interests as a higher priority than protection of the environment. The state's water resources behemoth, the Massachusetts Water Resources Authority, which serves metropolitan Boston and has a big influence on state water policy, might be considered the area's water god.

Other individuals and groups that wield control and power over water across the country include some water lawyers and analysts.

Historical Precedent

The concept of individuals or powerful groups controlling our water isn't a new phenomenon. "What people need to realize is this practice has been going on for a long time," says Colorado and

Rio Grande river basins expert Steve Mumme. "The pattern was set, or at least metaphorically set, by the Los Angeles water district when it started quietly buying up the Owens Valley and everything in between ... [William] Mulholland was the ultimate Southern California water czar," Mumme adds.

Behind-the-Scenes Agendas

When the Los Angeles Water Department bought up land along the Owens River in the early twentieth century, ostensibly for a new irrigation system for Owens Valley, reality wasn't as it seemed. Not much has changed since then.

The real motivation—positive or negative—behind the actions of various water rulers of today may not be obvious, either. Remember

---

WATER TALES

 What happens when your neighbors' wells run dry? A rancher in rural suburban Denver found out the pocketbook effect firsthand when he wanted to sell off some of his land for development. He had potential buyers, too. The only problem was that even though he owned the rights to plenty of surface water on his property—Colorado uses a prior appropriation approach to water—the local water board would not allow him to use those water rights for the development. Instead, he was forced to join the local water district and pay approximately $200,000 for water taps for the properties he planned to sell.

That water district was created when the aquifer level dropped and many of the surrounding property owners' wells ran dry. The district needed the water-rich rancher's participation, not for his water, but for his cash to help pay for the neighborhood's water and water infrastructure.

Footnote: Today the rancher still owns plenty of taps, and pays his share of the district's infrastructure costs, but there isn't enough water available to provide him with the full allocation he's entitled to in accordance with the water rights he owns! Such is the world of water in the arid Southwest.

the Tampa Bay Water Wars that dragged on? Then one day, the gods tempered their stances and agreed to compromise, or so it seemed to outsiders. Perhaps the water war and those who orchestrated it—read that "warred over the issues"—had another real strategy. Says Orlando water attorney Roger Sims, "There's a strategy of fighting and putting off the inevitable and saving a lot of money by not investing in alternative supplies any sooner than you must. I think that was part of the deal. At least that's the rumor."

Sound a bit fishy? Unfortunately, it's not. It's part of the reality swirling around water no matter where you live.

T. Boone Pickens, Texas billionaire oilman and energy czar, could be another of the water elite. He's touting an alternative energy plan that utilizes wind for power. It's a great idea, but some think Pickens' motives go well beyond energy. He's also head of Mesa Water, a small company in the Texas Panhandle with big plans to transport hundreds of thousands of acre-feet—tens of billions of gallons—of water via pipeline from the Ogallala aquifer to parched north-central Texas or San Antonio. Mesa Water's web site proclaims, "Mesa Water is ready to sell water to communities that don't have enough for the future" (http://mesawater.com).

WATER AND LAND DEVELOPMENT

Water is a simple formula: H_2O equals two molecules of hydrogen and one of oxygen. However, its disbursement can be a complicated equation. And, where there's complicated confusion, there are savvy investors who have learned to capitalize. One of the companies that retired tax attorney Robert A. Lembke helms is The Bromley Companies, LLC, a Denver-based real estate developer and water seller. "A key factor in developing land is water," the company touts on its web site. Lembke and his company develop and sell property along Colorado's Front Range, which includes metropolitan Denver. He's on a number of boards involved in water and, according to his web site, wields water power with more than 5,000 acre-feet of water (more than 1.6 billion gallons) in an often-parched landscape. For developers large or small with a need for water, he's the go-to guy in an area where the resource and rights to it can be scarce. Of course, the water comes at a price. "At least for Colorado, we have an infrastructure and distribution problem more than we have a water problem," says Lembke. "In most areas of the country—other than perhaps Nevada and parts of Arizona—we really don't have a water shortage. We have a thinking shortage. Usually that's a distribution or planning issue, and they're significant. That's where the failures have been, and now you have some really dry areas where the problems are even bigger."

Lembke and his associates developed land for commercial and residential use in Adams and Weld counties in Colorado, and in the process created a water system to handle others' water needs, too. That *system* is the United Water and Sanitation District. Lembke says it's a "facilitating entity" that provides water to other districts and owns reservoirs and an underground storage facility. United Water actually is a one-acre parcel of land that's been designated a "special district," which gives it the legal right to condemn land and to buy and sell water rights in what is a perennially very thirsty state.

Don't think Lembke and his associates are one of a kind or that they're a phenomenon particular to Colorado and the arid West. There are hundreds more public and private Lembke-types who have learned to use the laws to their advantage and are poised to capitalize in the water world. (Pickens capitalized on the special district approach, too.)

GOVERNMENT'S ROLE

Remember New Berlin, Wisconsin? The water god that city needs to please is the Michigan Department of Natural Resources. It wields the power and decides whether New Berlin gets its water. It did. The water gods' pantheon includes plenty of government entities—state and regional governments as well as quasi-governmental bodies like some river basin commissions and their leaders and boards, state engineers, and other officials charged with managing water in their states. Let's look at a few.

Water Management Districts

Water management districts take all shapes and forms, from citizen advocacy, oversight, and lobbying entities like the Charles River

WATER TALES

You be the judge in a battle that pits the city of Detroit, Michigan, and its Detroit Water and Sewerage Department against the Ontario (Canada) Ministry of Environment. In December 2008, the latter accused Detroit of "stealing" hundreds of billions of gallons of "Canadian" water over the last 44 years. It seems that a major intake into the Detroit River for Detroit's water supply is about 100 yards over the Canadian side of the border. Ontario wants part of its water put back, plus other restrictions and controls. In March 2009, Ontario granted the city of Detroit an exemption to "take" the water, with a cap on the amount.

Joseph Dellapenna, the Philadelphia-based water law expert and professor of law, offers this opinion on the scenario:

> Ontario is saying that every drop of water those eastern Michigan communities have been consuming since Day One is stolen Canadian water and it wants to be compensated. That supposes that water stays put the way land more or less does. If I build my driveway and it's two feet across the property boundary line, that's clearly on your property and it's clearly taking your property. But if I put my pipe into the creek and it ends up somehow on your side of the boundary line, whose water am I taking? Even if I'm taking your water, some of my water is going to flow over to replace your water.
>
> To say all that water is stolen Canadian water is utter nonsense. That completely ignores the reality of hydrology. But it's the way most people think because most people haven't thought seriously about water and the nature of water, about its being inherently a shared resource, and what that tells us about water law, water rights, water duties, and more.
>
> Looking at it another way, this claim would make sense only if taking water was like digging sand out of a sand pit—the sand comes from a pit, and a hole remains. We know, of course, that water will flow in to fill the "hole," so that some of the water coming into the Detroit [River] (regardless of which side of the border) comes from Michigan and some from

Ontario. In fact, the water is a shared resource and not simply a resource of either Ontario or Michigan. Such a claim illustrates the simplistic (or unsophisticated) thinking that all too often passes for analysis of water issues—particularly water

Watershed Association in Massachusetts to government agencies that have both power and financial backing. South Florida Water Management District is one of the latter, a regional governmental agency that handles water quality, flood control, water supply (distribution), and environmental restoration in sixteen Florida counties from Orlando in central Florida to the Florida Keys in the south. Members of its governing board are the ones to humor if anyone wants water in their jurisdiction. That principle applies whether it's the Old Collier Golf Club in Naples, which in 2008 went looking for a water-shortage variance (an exception to water restrictions) to keep its landscape flourishing (they got it),[1] or the city of St. Cloud and Orange County, which in 2007 sought a use permit to withdraw surface water from East Lake Tohopekaliga in the Kissimmee Chain of Lakes to meet "existing and future urban irrigation demands" (they didn't get it). But stay tuned. That battle isn't over. After appealing the ruling, St. Cloud did begin a water-use plan project to find alternative water sources, funded in part with a hefty contribution from the district.[2]

SOUTH FLORIDA WATER MANAGEMENT DISTRICT

The South Florida Water Management District's governing board also has led efforts to restore Florida's Everglades. In December 2008, the district approved a $1.34 billion purchase of a minimum 180,000 acres in South Florida from U.S. Sugar Corporation in an effort to stop pollution of the Everglades. But the deal ran up against budget constraints. In May 2009, the deal had shrunk to 72,500 acres and $530 million with the option to purchase more acreage later, and by August 2010, it shrunk again, this time to $197 million for 26,800 acres with options to buy more.[3]

Federal Participants

Elsewhere in the East, the U.S. Army Corp of Engineers has a huge say in how water is managed, says Philadelphia water attorney Kenneth J. Warren. However, unlike in the West where appropriated water rights can limit government discretion, in the East government agencies make the major water allocation decisions.

"Private individuals are not making these major water decisions because most of the Eastern states have a permitting system applicable to water withdrawals within the state, and when interstate conflicts arise, compacts and Supreme Court decisions have in some regions resolved the disputes," says Warren.

That doesn't make water decisions any less complicated. Take the Delaware River Basin, which drains parts of Delaware, Pennsylvania, New York, and New Jersey. Engineering representatives from the environmental departments of the various states involved, as well as from New York City, play a huge role in managing water, Warren says. This system exists because in a 1954 decree, the U.S. Supreme Court allocated Delaware Basin water among the basin states and New York City. The Delaware River Basin Compact, which established the Delaware River Basin Commission, gave the commission the right to allocate water resources as long as it does not adversely affect the rights of the parties under the Supreme Court decree. New York City and the four states in the basin are parties to the Supreme Court decree, and as such each essentially has a veto in cases where a water allocation might affect their rights under the Supreme Court ruling, says Warren. (Figure that one out!)

Water Resource Directors

In Pennsylvania, a state water director might be a good idea, says Dellapenna. Many individuals play a role in deciding who gets the water, but no one at the state level regulates how water is used, he adds. That's problematic in a state that straddles three major watersheds—the Delaware River, the Susquehanna River, and the Ohio River watersheds—plus some minor ones.

As a result, a statewide water crisis is rare. "One watershed may be in crisis and someone will say we need water-law reform, but two-thirds of the state isn't and says, 'No, we don't.' If someone started talking about inter-basin transfers, we might actually get water-law reform," says Dellapenna.

Head west about 2,500 miles to the state of Idaho, and it's the director of the Idaho Department of Water Resources who is figuratively and literally charged with turning on, or off, the tap. Idaho law requires that the director administer water rights under the prior appropriation doctrine. That means cutting the water off to junior water rights holders if and when it's necessary. It's the law of the land.

The Role of the Courts

The great mitigator when it comes to the water elite is the courts, says Bruce Newcomb. He's former speaker of the Idaho Legislature (1998–2006), a longtime Idaho rancher who uses irrigation, and director of government affairs for Boise State University.

"If anyone was close to being or trying to be the water god in Idaho, it was the Idaho Power Company," says Newcomb. "It hasn't been able to achieve that status, largely due to the court system."

People with money and political power who are accustomed to having their way may think they can become part of the water elite and everyone must kowtow to them. "The entity that makes sure that doesn't happen is the court system," Newcomb says. "In the end, it prevents many such abuses."

The drawback of water cases ending up in court is the cost. "You better have a truckload or a trainload—at least a couple of cars full— of money," says Newcomb, who has joined with other ranchers to form a water district. "That's why I belong to this water district. It's economies of scale. Individually I couldn't do what needs to be done to resolve water disputes. But together we become a strong enough entity with enough resources behind us that we can stand up to anybody."

NATIONAL WATER CHIEF

Given the pervasiveness and complexity of water ownership and the multitude of directors, controllers, buyers, sellers, brokers, lawyers, and others involved, it makes sense that the United States would have a big kahuna of water. Who is it? The short answer may surprise you. No one. Although such a post has been considered, and a number of groups have urged the Obama administration to appoint a national water czar, as of press time no appointment had been made. However, the administration has made a number of water-related appointments—mini-czars, including:

- Deputy Interior Secretary David J. Hayes as the water czar for California
- Cameron Davis, then president and CEO of the Alliance for the Great Lakes, as special adviser to the U.S. EPA overseeing its Great Lakes restoration plan
- John Tubbs, a long-time Montana water resources administrator, as deputy assistant secretary for water and science
- Anne Castle, prominent Denver water attorney, as assistant secretary of the interior for water and science

(The United Nations appointed a "water czar," Maude Barlow of Canada, in December 2008.)

The longer answer, to the U.S. water czar question, however—as we discussed in Chapter 5—involves politics and power as well as personal, state, regional, and local agendas.

Policy for All

In 2007, New Mexico governor Bill Richardson—then a Democratic presidential candidate—floated a proposal to develop a national water policy and create a U.S. Department of Water that would be headed by a cabinet-level secretary of water. That policy would include the sharing of resources between those states with water—including those on the Great Lakes—and those without. The outrage against the proposal and Richardson was immediate.

"If you want to touch off fireworks for the next millennium, push that concept of national control of water," says Roger Sims, past chair of the Florida Bar Environmental and Land Use Law Section and member of the American Bar Association Standing Committee on Environmental Law. "A state water board was proposed in Florida, and the backlash was unbelievable. Those areas that have water don't want anyone managing their water out from under them. And those that don't have it obviously would like some help from state or federal government to make water from water-rich areas available to them."

Idaho rancher and former legislator Newcomb doesn't like the idea of a national water policy, either. "I think that's very, very dangerous. It's a states' issue," he says. "Here is Idaho sitting with two representatives in Congress, and California with fifty-two representatives. So you're going to establish water policy on a federal level when you're outgunned?"

Newcomb argues that water policy would then become a political issue rather than a question of public policy. California would be able to take Idaho's water, he says, "because fifty-two representatives have a lot more clout than two. Plus, if you look at the seniority system and leadership in Congress, Idaho basically loses on that."

Newcomb fears that the more populated states would rule when it comes to policy, and coalitions of states like California and Nevada, which both need water, could gang up politically on states like Idaho. "That's not the way it should be, big against little," Newcomb adds.

As for the possibility of centralized control in the future, Sims is doubtful. "Some areas have concentrations of people who want water and have votes and vote for politicians who want the same," he says. "Other areas aren't as well populated and have a different agenda, and they don't have as many voters."

Given that so many disputes over water are between states, it might make sense to have a national policy or law to regulate interstate water issues, says Dellapenna. However, he agrees that isn't going to happen. "Congress theoretically has the power, but it's never done it and it's not likely to do it because of political reality."

Some things Congress theoretically could do to address water issues, says Dellapenna, include naming a secretary of water; giving the U.S. Department of Interior the power to allocate water among states, and setting up a special set of courts to settle water disputes.

East or West, water experts are doubtful about the chances anytime soon of a national water god. The idea of a national water policy is "misguided," says Clive Strong, chief of Idaho's Natural Resources Division. "A national water policy ignores the fact that in large part, every water basin has different hydrologic characteristics, and therefore water allocation must be tailored to the characteristics of each basin. For instance, a policy that fits the Snake River Basin would not work well in the Colorado River Basin," he says.

Given those hydrologic differences, it's important that states develop their own water policies, says Strong. "That's not to say there isn't a federal role to be played. Under the Clean Water Act, the federal government has established certain minimum standards between states in terms of water quality."

"It's a matter of thinking of the country as a connected whole that uses energy and water resources independently," says Daniel Williams, Seattle-based urban design and sustainability expert. "So, even though each state has its own way of doing things, everybody must work together on this because it's the United States of America. A national advisory board on water makes practical sense," says the Sierra Club's Ken Kramer. Bringing together experts and approaches from different parts of the country could provide solutions to the overall approach to water and how to respond to water issues nationwide. It's feasible, too, says Kramer.

Williams agrees on the national aspect of the solution. "We must establish a national dialogue that says, 'Since the beginning of civilization, we've acted as if there will always be enough resources. But we're also students of civilization and we've seen that huge advanced civilizations go belly up because they've run out of resources—typically, water.'"

Because we're in a similar position that's likely to worsen, Williams says, we need a 100-year plan on a very large scale—state- and continent-wide—to ensure we have enough water to sustain ourselves "rather than spend an incredible amount of money, which means taxes, to ship water from Point A to Point B."

Water is a top issue for the twenty-first century. "We have to be more innovative and aggressive in how we deal with water issues," Kramer adds. "There is a growing awareness of that and I doubt there would be political opposition to a national advisory board. The key is *advisory*." But Kramer also predicts strong political opposition to any attempts at national control of water decisions.

The first step, rather than taking away states' rights, is to take an architect's or planner's approach to the problem, says Williams. Bring people—at least one person from every state and key federal, state, and scientific agency—together on a board, and then look at the entire United States and establish what we have. Where are the best soils, for example, or the best water supplies? The best aquifers? The best water storage area? The flattest land? And so on.

This board has to have huge teeth, and start with the premise that as a nation we're out of water. "This is not business as usual. This is not about who gets their spade in first or who gets the first bucket of water," says Williams. "We're in this to look at future patterns of growth and development. . . . [We can no longer afford] to build where there is no water, where there is no good crop-growing soil, and where there are no roads, and then have to go through a tremendous expense to provide infrastructure to make that happen. We need to change our pumps-and-pipes engineering approach and make it more of an ecological planning approach. You can't do it with brute muscle. There's a dance we need to learn."

Floating the Idea

The issue of national water policy came up in 2003 at a legislative hearing of the House Committee on Natural Resources' Subcommittee on Water and Power at which Peter Gleick of the Pacific Institute suggested setting up a national water commission:[4]

> It is time for a new national water commission. The Pacific Institute has called for the creation of a National Commission on Water for the Twenty-First Century to provide guidance and direction on the appropriate role of the United States in addressing national and international water issues. The Commission must be nonpartisan and include representation from across the many disciplines affected, including the sciences, economics, public policy, law, governments, public interest groups, and appropriate private sectors. While the duration of the commission should be fixed, adequate financial resources should be provided to permit it to do a serious and effective job.

"We haven't had a serious national effort to analyze water policy nationwide since the old National Water Commission folded its doors in the 1970s," says Colorado State University's Steve Mumme.

That may (or may not) be changing, however, as talk of a national water policy persists. Meanwhile, the nation's lawmakers deal with some of today's water issues, and introduce, talk about, and refer to committee (or something else entirely) various related bills. Some flounder; others become law. The latter, for example, include the Secure and Responsible Drug Disposal Act of 2010 (establishing programs for safe disposal of unwanted or old prescription drugs and controlled substances), and several bills approved by the lame duck Congress in December 2010 and subsequently signed by President Obama in early January 2011. They include the Reduction of Lead in Drinking Water Act (amending the Safe Drinking Water Act to cut lead in drinking water), the Claims Resolution Act of 2010 (addresses, among other things, Native American compensation related to water rights), and an amendment to the Clean Water Act that allows for federal agencies to pay water pollution and control fees.

Sandia National Laboratories' Mike Hightower points to Israel as an example of how managing water at a national level can work even in a country that once was mostly desert. "Everything is managed nationally for the best options for the whole country," says Hightower. "It works in Israel because it's a smaller country. But you are seeing much more interest in improving water management and planning on a much larger scale than we do in the United States. I think there are opportunities for organizations to cooperate and improve water resources planning, resources management, and resources allocation."

Water policy is considered the prerogative of the states, adds Mumme, but the federal government does have some piecemeal involvement. For example, it is involved where states share watersheds; in the Colorado and Rio Grande river basins; in funding and financing infrastructure for reclamation works in the western United States; and through federal reserved water rights for Native Americans arising from the *Winters* Doctrine.

Developing a national water policy is very important, says Kramer. "I do think, however, a national policy must be based on pushing for general principles that really apply nationwide, like providing incentives for water conservation and water efficiency," he says.

"The focus should be on these types of guidelines, not water allocation, because [since] each state has its own system for regulating the withdrawal of surface water and groundwater, it would be a mistake to push for a national system that would standardize withdrawal among all fifty states," says Kramer. "For one thing, it's politically impractical, and secondly, it would cause so much disruption in water systems around the country that it wouldn't be worth the effort."

Instead, Kramer suggests changes such as improving established national plumbing standards with a goal of greater water-use efficiency across all states. Low-water flush toilets are a good example. Some states may want to promote ultra-efficient 1.1-gallon flush toilets as compared with the national low-water 1.6-gallon standard. "Right now, if a state or community wants to institute a standard that's more aggressive than the national plumbing standard, they're not preempted from doing that. But there's no incentive to do it, either," he adds.

Getting consensus on water is a tough task. Mumme, also a well-known international water treaty negotiator, recounts how difficult it can be:

> In 2002, I chaired a water summit in El Paso sponsored by Representative Sylvester Rojas's 16th District in Texas. Rojas had been perennially frustrated by the difficulty in getting all the parties on the same page to target or identify legislative measures that could bring some coherence and rationality to the use of water on the Rio Grande and in the groundwater basins associated with it.
>
> My argument was that a smart move would be to have some sort of Mexico and United States agreement with a definition or set of warning signs about pending water scarcity in the river based on climate, water usage, and a variety of other things. That way, if an alert on the river were issued, all the different users who rely on the river's water could curtail their use for a set time until the river was out of the stress zone.
>
> We couldn't even get an agreement on that because there's so much competition for the water resources that different users feel they're going to lose if they don't control the process.
>
> Right now, we're planning by disaster. I don't want to be an alarmist, but I think that disaster is fairly imminent.

A New Thought Process

Las Vegas's Pat Mulroy, one of the country's oft-mentioned water elite, suggests a fundamental rethinking of the nation's water policy, but with a twist. "There are areas that need water quality protection. There are areas that need flood protection. And there are areas that need drought protection," says Mulroy. "You don't need a national water czar. I would create a water coordinating council that would bring together the Army Corps of Engineers, the Environmental Protection Agency, the Department of the Interior, and maybe a water quality assessment group, and have a national coordinating committee—almost like an interdisciplinary team."

In the vein of cooperation, Mulroy says that her agency, the Southern Nevada Water Authority, has teamed with a number of other major water utilities, including those in Seattle, Portland, San Francisco, San Diego, Denver, and New York, to form a climate action committee. The group is working toward creating a cohesive research effort to come up with a central global climate model. That's important, says Mulroy, because the real potential of a water crisis relates to climate change, and currently there are 20 different climate models.

"The entire dialogue around climate change has been about mitigating carbon emission and combating the problem of global warming. What we haven't talked about is this broader problem. Some consequences of global warming already are beginning to play themselves out, and a strategy has to be developed—especially for water utilities—on how to adapt and what the possible consequences of climate change are in various areas of the country," Mulroy adds.

WATER REALITIES

- Those who wield water power range from state engineers and directors of state departments of natural resources to heads of water utilities, watershed organizations, and regional water boards with power and mission, to private individuals capitalizing on their large water rights holdings, and more.

- Water management districts take all shapes and forms, from those focused on citizen advocacy, oversight, and lobbying, like the Charles River Watershed Association in Massachusetts, to government agencies that wield plenty of influence and have financial backing.
- The federal government weighs in as god, too. The U.S. Bureau of Reclamation and the Army Corps of Engineers have a huge say in how water is managed, along with other government entities, representatives, and employees.
- Some experts urge creation of national water policies or advisory boards with regional and local approaches to help solve national, regional, and local water issues.
- Other experts suggest that taking control of water away from states may be a political impossibility.
- The only consensus is that the United States has big water problems and needs to find solutions.

NOTES

1. South Florida Water Management District, "In the Matter of Water Shortage Variance Application No. 4386 Filed by Old Collier Golf Club," Order Granting Variance, February 27, 2008, https://my.sfwmd.gov/pls/portal/docs/PAGE/COMMON/PDF/SPLASH/2008_087_OLDCOLLIERGOLF_VAR.PDF.
2. *City of St. Cloud vs. South Florida Water Management District*, Settlement Agreement (September 4, 2008), http://my.sfwmd.gov; South Florida Water Management District, press releases, "Water Managers, St. Cloud, Orange County Commissioners Resolve Legal Challenge"; "SFWMD Strengthens Public Assurances and Financial Safeguards for Land Purchases," September 11, 2008, https://my.sfwmd.gov/portal/page?_pageid=3034,21168995&_dad=portal&_schema=PORTAL.
3. South Florida Water Management District, press release, "SFWMD Governing Board Takes Next Step Toward Historic Land Acquisition for Everglades Restoration," December 30, 2008, https://my.sfwmd.gov/portal/page?_pageid=3034,21169019&_dad=portal&_schema=PORTAL; Florida Department of Environmental Protection, press release, "Governor Crist Encourages Federal Support for Everglades Restoration and Apalachicola River and Bay,"

May 28, 2009, http://dep.state.fl.us/secretary/news/2009/05/0528_02.htm; SFWMD, "Reviving the River of Grass ... August 2010," https://my.sfwmd .gov/portal/page/portal/pg_grp_sfwmd_koe/pg_sfwmd_koe_riverofgrass.

4. Testimony of Dr. Peter H. Gleick before the Legislative Hearing of the Subcommittee on Water and Power of the Committee on Natural Resources, U.S. Congress, April 1, 2003, http://pacinst.org/publications/testimony/ national_water_commission.htm.

7

THE COST OF WATER

CASH, COMMODITIES, AND CAPITALISM

When the well runs dry, they know the worth of water.
— Benjamin Franklin

The cost of water—whether from the tap, bottle, or irrigation ditch—is rising and will keep doing so as supplies and availability shrink. When was the last time you bought a cup of coffee or a bottle of water? Chances are the water cost more than the coffee did. That 16-ounce bottle of Evian easily could set you back close to two bucks at the convenience store. (In some states, you'll need to add another several cents for the plastic bottle deposit.) A simple cup of joe, on the other hand, could cost $1 or less, no deposit required. Even a latte at $4 is a bargain compared with a designer bottle of water, which easily can run $8 and up.

That's a frightening thought, considering that 70 percent of Earth's surface is water and, theoretically at least, anyone can put out a pot to collect rainwater free of charge (if it's legal in your state). These days, though, there's no guarantee on the quality of what falls from the sky!

As supplies of safe drinking water become scarcer, and as science and medicine reveal the dangers of additives in water supplies, more people willingly—or unwillingly—pay a premium for their water, at least the bottled variety. Even people who drink tap water pay more, especially after factoring in the effect of shrinking water supplies on the costs of producing food and raw materials. Throw in the ever-increasing billions of dollars spent annually to keep tap water safe and to update antiquated infrastructure, and tap water suddenly isn't so cheap.

Whenever a commodity is in demand, you can count on cash changing hands and capitalism springing up. Climate change and environmental issues aside, there's little question that water is fast becoming the next great commodity of the twenty-first century. Markets are hot, and the controversies are hotter. Let's look more closely.

FACTS AND FIGURES

- The numbers tell the story of water's importance. We've briefly mentioned how much water Americans use, but it's worth reviewing it with capitalism in mind. The United States uses 410 billion gallons of water every day, 349 billion gallons of which are freshwater. That's nearly 150 *trillion* gallons of water a year.
- The average American uses about 100 gallons of water a day for in-home personal use, but that can vary depending on climate, location, and other factors.
- Americans consumed an estimated 8.1 billion gallons of bottled water in 2009, according to the Beverage Marketing Corporation.[1]

Dollars and Cents

Water is cheap, says Steve Maxwell, widely recognized water-business expert and managing director of TechKNOWLEDGEy Strategic Group, a Colorado-based independent investment banking and management-consulting firm that focuses on commercial water and environmental services industries. "But in many regards, the treatment, collection, transmission, and storage of water in terms of dollar costs represents one of the largest industries in the world. It costs billions of dollars a year to move this water around and particularly to treat it to levels suitable for drinking," says Maxwell, author of the upcoming book *The Future of Water* (American Water Works Association, 2011).

We're talking about astronomical amounts of money spent on water, far in excess of the hundreds of millions of dollars that change hands relative to water rights, adjudications, court cases, treaties, compacts, and water-related investments. The average residential monthly water bill likely is affordable (for now), but the dollars add up rapidly when you remember that everyone requires water.

Consider a few cost numbers:

- Residential customers in the United States paid an estimated $23 billion for their water in 2010, according to estimates from Raftelis Financial Consultants (RFC), based on numbers from their biennial survey, conducted with the major industry group American Water Works Association (AWWA).[2] That doesn't include billions more dollars spent by industrial and commercial users.
- Though prices vary dramatically depending on geographic location, rate structures, amounts, and other variables, the AWWA/RFC survey reports that the average monthly residential water charge was $27.82 for 1,000 cubic feet (7,480 gallons) in 2010, up 89.2 percent from 1996. The average monthly rate for a commercial or light industrial user for 50,000 cubic feet of water or (374,000 gallons) was $1,112.58, and for an industrial user 1 million cubic feet (7.48 million gallons) was $21,710.52.
- Americans spent an estimated nearly $10.6 billion on bottled water in 2009, according to the Beverage Marketing Corporation.

THE VALUE OF WATER

No matter how water is appropriated or who uses it, we can't live without it. We need water to drink, produce our food, create our power, manufacture and ship our goods, and carry on our everyday lives. Therefore, it should be priced accordingly. Right?

The answer depends on whom you talk to, where they live, whether they view water as a public or private resource, whether they have enough water, and so on. "Accordingly" varies dramatically, too. The price of water is yet one more dispute swirling around the nation's water supplies.

The debate over whether water is a public right or a private resource isn't an either-or question, says Maxwell. A business- or private capital–oriented thinker can't simply decide that if someone can't afford to pay for water, he or she doesn't get any. "Clearly everyone needs water to survive, and as a society we have to find some way to provide that," says Maxwell. "Likewise it's not necessarily rational to say that there should be no role for private industry or private capital in this business. If you look at the record of small, struggling, bankrupt municipalities trying to provide water, meet regulations, and stay up with technology, that's a pretty sad record, too. So obviously it has to be some balance in the middle."

It makes no economic sense, either, to have multiple water-distribution infrastructure systems in a single neighborhood, says Maxwell. "Water is clearly a natural monopoly that has to be regulated by some sort of oversight—which is the way it works for private utilities," says Maxwell. "There has to be a better way to treat water as a commodity and try to manage it as a commodity, utilizing market forces overseen by regulatory agencies. That's working in many parts of the world. It just hasn't worked very well here yet."

One of the challenges is that people perceive water as a resource that falls out of the sky and runs free. Why should they have to pay for it, they ask, and particularly why should they pay higher prices? That attitude, Maxwell says, stems in part from the massive federally funded—as opposed to locally funded—water projects that have made high-growth cities in the Southwestern deserts possible. Cities like Phoenix, Albuquerque, and Las Vegas would not be able to sustain their booming populations without these federal water projects.

"We are going to see more attempts to apply market forces to this business and to ensure that people are paying what it really does cost to get that water to them," says Maxwell. "In areas where it doesn't make a lot of sense to have huge concentrations of people—like in the high deserts—people will ultimately pay a lot more for their water."

In case you're considering putting a barrel in your yard to catch the water falling from the sky, it's best to check your state's laws. In some states, especially those with prior-appropriation water laws, you may not have the right to water you catch in a barrel on your property. That's because the water already is accounted for and the rights to it allocated somewhere else in the system.

The Price of Your Water

Water in the United States is a deal, even factoring in the growing number of utilities raising rates across the country. Though prices vary wildly, consider, for example, a water bill of $35 for 8,000 gallons of water per month. That equates to 228 gallons of treated and tested water per dollar—delivered and available to each home and business 24 hours a day, seven days a week. "Compared with other commodities that we rely on in our daily lives, it is a pretty good deal," says Greg Huff, CEO of the Iowa Rural Water Association, an industry organization that provides training and assistance to the Iowa water and wastewater industry.

Let's look at how much Americans paid for their water in 2010. Municipal water rates were highest in Atlanta, Georgia ($7.30 per thousand gallons), and lowest in Savannah, Georgia ($1.25 per thousand gallons), according to a 51-city survey by Park Ridge, New Jersey–based NUS Consulting Group, energy cost consultants. Though NUS looked only at commercial and industrial users in those cities, its survey provides a snapshot of cost trends. More results of their survey of 2010 and 2009 rates include:

- Los Angeles had the biggest year-over-year increase in water rates, up 20.7 percent to $5.02/1,000 gallons in 2010.
- 2010 rates remained unchanged in 13 of the 51 cities, from the East to the Midwest, to the South and Southwest.
- Among the 51 cities, the average cost per 1,000 gallons was $3.32 per month.
- No city dropped its rates in 2010 from 2009 levels.

Water rates are very much a local, as opposed to global, supply and demand issue, says Maxwell. That's an important consideration. Rates can be wildly divergent between two geographic areas.

That price gap applies to different countries, too, he says. "We pay a fraction of rates that people are used to paying in countries like Germany, France, and Holland."

WATER FACTS

A 51-city survey of municipal water rates by Park Ridge, New Jersey–based NUS Consulting Group* (http://nus consulting.com) showed the following water rate trends for 2009–2010:

Rates Unchanged	Double-Digit Increases
Huntington, West Virginia	Los Angeles, California: 20.7%
Pittsburgh, Pennsylvania	Providence, Rhode Island: 16.6%
Binghamton, New York	San Francisco, California: 16.1%
New Orleans, Louisiana	Chicago, Illinois: 14%
Trenton, New Jersey	New York City: 13%
Newark, New Jersey	Atlanta, Georgia: 12.5%
Dover, Delaware	Sioux Falls, South Dakota: 12.1%
Indianapolis, Indiana	Miami, Florida: 12.1%
Albuquerque, New Mexico	St. Louis, Missouri: 12.1%
Grand Forks, North Dakota	Portland, Oregon: 12%
El Paso, Texas	Houston, Texas: 11.4%
Green Bay, Wisconsin	Biloxi, Mississippi: 10.5%
Memphis, Tennessee	Burlington, Vermont: 10.1%
	Newport, New Hampshire: 10.1%

*NUS Consulting Group, "2009–2010 International Water Report & Cost Survey," abridged version (2010), http://nusconsulting.com.

Rate structures. The methods used to determine water rates vary across the country. Some rate structures charge storm-water or run-off fees, others don't. Certain areas don't meter water usage, and determine cost by usage amounts. Alternatively, consumers may pay a flat rate, no matter how much water—1,000 gallons or 100,000 gallons—they use during a given period. Other locales have tiered systems so that the more water one uses, the higher the price one pays per unit, whatever that unit amount is. Though increasingly the approach is disappearing, some municipalities—including in water-rich Iowa—still have declining rate structures: The more water you use, the less the cost, says Huff. This kind of a fee structure could be used as an incentive to attract water-intensive businesses to a community to spur economic development. But that's certainly not always the case. Experts, including Huff, agree that this approach to water must change and is changing. Otherwise, there's little incentive for people to consider conservation, which must be a cornerstone of twenty-first-century water philosophy.

Iowa isn't the only place where water usage pricing may not reflect national water concerns. Many cities around the United States don't meter water, adds Maxwell. "Those approaches may have worked 50 years ago or in an area not yet at a water-crisis level today. But they certainly won't work in the future," says Maxwell. The trend is toward much more accurate monitoring of water usage, whether at a residential level or an agricultural one.

There's also little doubt that water rates, even in water-rich states, will eventually go up. The questions become when and by how much. The answers will depend on supply, demand, geographic location, and more.

Water rates will continue to escalate for several reasons, says Peiffer Brandt, RFC's project manager for the AWWA/RFC biennial rate survey. These include aging infrastructure, increasing regulations, and additional source-of-supply costs. Historically, the cost of providing water has been subsidized, says Brandt. "There has been the belief that availability of water is almost a God-given right. We've shifted from that viewpoint, and are evolving to the point where rates will reflect the full cost of providing the service."

The price of water still isn't high enough to jolt consumers, no matter where they live, says Southwestern water expert Steve Mumme. "Whether you're talking about farmers consuming water for production or folks watering their lawns, or water for amenities in the cities, we're not sending the right messages to people about . . . our [shrinking] water supply," he says.

The Price/Value Chasm

Although many experts agree that water is undervalued across the United States, the reasons for it vary.

"It costs more to have cell phone service than it does to get water," adds Sandy Gaines, water resources expert and director of the Utton Transboundary Resources Center at the University of New Mexico.

The public health equation. "As long as people turn on that tap and water runs, it's an 'out of sight, out of mind' issue for many," says Huff. "It's similar to electricity—as long as it's on, you pay your bills, and don't think much about it. The difference is that water is a necessity of life. It's a public health issue around the world. When we turn on our tap, we don't worry that we're going to get dysentery or cholera or anything else from our water. . . . But do we value that?"

Obviously, says Huff, we all want safe drinking water, so we think of it to some extent. But do we see the big picture and consider the water supply globally? For instance, he asks, "Do we look at U.N. statistics as to how many children die every day from waterborne diseases? We don't really think about how vital public water systems are in the United States to maintaining public health."

Haves and have-nots. Utilities have done too good a job in providing a very good service at a very low price, says El Paso's Ed Archuleta. "The price of water has to be valued and recognized. The cheap water is gone, and people will have to pay more in the future."

Archuleta, like many water officials across the country, advocates public education initiatives to help consumers better understand their community's water system. To that end, many utilities provide education programs that start in preschools and actively promote water awareness and conservation at all levels.

Why not just raise rates? If water is so undervalued and underpriced, why not simply raise the rates? That's not necessarily the best approach in the short term or across the board, experts agree.

People don't seem to mind paying for a bottle of water, but when you raise their water bill $1 or $2, the wheels come off, says Archuleta.

Pricing varies widely and is hotly debated, says Jeffrey Kightlinger, general manager of the Metropolitan Water District of Southern California, a water wholesaler. "Many economists say water is significantly undervalued in the way we price it," says Kightlinger. "Typically, we charge customers the cost of delivering the water. The price of water then is structured to cover our costs, and you see it vary widely as a result. Some agriculture districts pay $15 to $20 an acre-foot, whereas in our service area, Metropolitan's wholesale price is $500 an acre-foot of water. By the time water gets to the homeowner through the retailer, it could be up to $1,000 per acre-foot," adds Kightlinger.

It's not that easy to raise rates, agrees Richard Atwater, Chino, California, water utility chief. Municipal water suppliers are public agencies, and that means rate changes must be approved by city councils and elected water board directors, who must consider individual hardships and economics.

"You have to be smart about raising rates," Atwater says. "For example, a lifeline rate for people like senior citizens who don't use a lot of water and perhaps can't afford higher rates is appropriate in most communities. But if you live on a large lot and want to use 10 times the normal amount of water to grow a gorgeous garden, you should have a much higher rate since it will cost more for the water utility to provide that additional water."

Catch-22. Beware, however; cheap water is a catch-22, says Pacific Institute's Peter Gleick. At a time when water conservation is important, too-cheap water promotes inefficient and wasteful use of the resource. Systems continue to leak, and people aren't overly concerned. That approach needs to change, Gleick adds.

Going forward, we can either find new sources or do more with what we have, says Gleick. "It's much cheaper and better for the environment to do more with the water we have. We can't keep finding new sources. There is a limit, and we're up against the limit now in many parts of the United States, and in many parts of the world."

One of those areas up against the limit is California. There is no silver bullet to solve that state's water problems, Gleick testified in January 2010 before a U.S. House Subcommittee on Water and Power. But, he also said at the time, "Improving the efficiency of our water use is the cheapest, easiest, fastest, and least destructive way to meet California's current and future water supply needs."[3]

How much "extra" water is lost through inefficiencies? In September, Gleick's organization, the Pacific Institute, released a report, "California's Next Million Acre-Feet: Saving Water, Energy, and Money," that identifies water savings totaling more than 1 million acre-feet of water a year. That's enough water, the report says, to satisfy the annual water needs of the city of San Francisco 12 times over![4]

Historically, the price of water has been very low, and because of that, people turn on the faucet without a second thought, says Gaines, a water resources expert who has focused on the Colorado and Rio Grande river basins. That's about to change, he says, and given current and future water supplies, climate change, and other factors, "we really do need to think about water."

The Infrastructure Cost Dilemma

With such inexpensive water, public utilities often can't afford to repair crumbling infrastructure, either. Inevitably, though, they will have to.

There are big unknowns with the infrastructure. "I don't believe the industry has fully grasped the issues and costs yet," says analyst Brandt. No one really knows how long the existing infrastructure will last or what the true cost to replace it will be. Today there often are streets, buildings, and other improvements above the under-ground infrastructure, so it will be much more expensive to replace than it was to originally install."

That infrastructure crisis already is being felt across the country. It's especially straining on small rural communities, like Humeston, Iowa, that have declining populations and increased—meaning costly—rules and regulations. "We're talking about the unseen infrastructure that carries drinking water supplies," says Huff. "Most people in a town think of their water system as the water tower they see. But most of the infrastructure is underground.

"As long as water comes out of the tap, things are okay," he adds. "But there's a ticking time bomb underground. In many places, water pipes have been in the ground for fifty to eighty years and longer. At some point that infrastructure will need massive replacement, which will hit not only the large communities in the pocketbook, but the small ones as well."

WATER TALES

A few years ago, the City Council in Humeston, Iowa, population 543, realized it absolutely had to replace the town's crumbling water system, which had been built in the early 1900s. Greg Huff, then a council member, says the pipes were so clogged with mineral buildup that they looked as if they had been buried in the ocean for years. However, the average person doesn't think of (or wouldn't know) that, because water systems are largely unseen and don't get dug up unless there's an interruption in service.

To help convince citizens of the tiny town that they really were facing a water crisis, officials displayed a somewhat frightening section of that worn-out and clogged pipe at City Hall.

The tactic worked. With the help of federal community block grants and other federal money, the small system was replaced to the tune of $1.1 million in 2003.

Not Always Too Little Water

Water-rich Idaho—a major exporter of the commodity—has its own infrastructure problems, which stretch far beyond that state's borders. "We don't have enough water storage or infrastructure like canals, waterways, and conveyance systems," says Idaho water resources expert David Tuthill.

The state is in the enviable position of having plenty of water. It stores about 16 million acre-feet annually, and more than 90 million acre-feet flow from the state every year. However, says Tuthill, that water isn't captured because the infrastructure—as in storage capacity—is inadequate. "We've added a number of uses that weren't contemplated when we built our infrastructure."

Those new uses include requirements that a certain amount of water flow through the rivers to protect endangered species and provide for fisheries habitat. "Even in a dry year, we'll send nearly a half-million acre-feet downstream in the Snake River system," says Tuthill.

Expanding urbanization and irrigation tax the system, too. Out of about 3.5 million acres irrigated in Idaho, about a million acres are irrigated from the Eastern Snake Plain aquifer, with diversion of about 2 million acre-feet of groundwater every year. Added energy uses, which demand surface water and groundwater, further strain Idaho's ability to satisfy existing needs, as do changes in snowpack levels and spring runoffs associated with climate change. In fact, says Tuthill, the biggest water issue in the United States today is climate change, and it's exacerbated by inadequate infrastructure. It's of the most concern in the Western states because they rely on melted snowpack to fill their reservoirs. If the snowpack is lower and therefore melts more quickly and easily, the infrastructure isn't in place to capture all the water. So even if the total supply of water remains the same, what's captured in the reservoirs won't be enough to meet water needs during the summer, Tuthill adds.

Despite all the parched fields, farmers versus fisheries battles, and drought concerns coming out of California, lack of water isn't the issue for that state, says Kightlinger. Developing major new infrastructure to harness, store, and direct that water is the problem. Southern California gets its water supply from three basic sources, he says: local supplies, the Colorado River, and Northern California. California is not a truly water-poor state. However, it does require the best use of the resource and good infrastructure to move it where it's needed.

"We have put off infrastructure investment for some time in this country, and that is having an impact on most utilities—roads, transportation, rail, airlines. It's impacting water supply as well," says Kightlinger. "We have a state [water] crisis because California has not invested in the conveyances it needs to move the water from water-rich Northern California to parched Southern California."

Cashing In on Agriculture versus Urban Use

The controversy swirling around pricing water commensurate with its value includes politics and personal agendas, too. One of the most contentious battles pits agriculture and irrigation interests against urban municipal and industrial users. Remember that in Chapter 1 we said that agricultural irrigation is the nation's second-biggest water user, accounting for some 128 billion gallons of freshwater every year, according to numbers from the U.S. Geological Survey.

WATER TALES

Cheap water, or, as in the case of Reading, Massachusetts, free water, helped to perpetuate what was an annual summer event for decades— the drying up of portions of Massachusetts' Ipswich River, according to Kerry Mackin, director of the Ipswich River Watershed Association. The town's municipal wells were located along the river's banks. There was never any cost to the town for the water itself— only the costs associated with pumping, treating, and transporting the water. When the town faced the high cost of building a new treatment plant it opted to stop groundwater pumping and instead turn to the Massachusetts Water Resources Authority, which serves metropolitan Boston, for its water supply.

Politics and pressure. Despite using so much water, agriculture is a small part of the U.S. economy. Nonetheless, it has broad political and social importance, says Gaines. "The use of water in cities gives you more economic value, but any attempt to transfer water away from agriculture and give it to the cities is politically controversial. If Denver or Las Vegas or Los Angeles needs more water, your first thought might be to see if farmers could use less. It's a logical scenario, and from a cold-hearted economic analysis, it would be a good move. But people and politicians are very reluctant to make those transfers on a large scale," Gaines adds.

Longtime Idaho rancher and former politician Bruce Newcomb also recognizes the dilemma. Exacerbating the problem, he says, is that some people feel they can easily turn to Third World countries like Mexico to provide food more cheaply. Those people think the better use of water is to provide it to large consumers like cities, technology manufacturers, and so forth. "But food safety then becomes a question, and we all realize that now," says Newcomb. "That's where the tensions come from."

As the resource becomes scarcer, it could be time to rethink how the nation uses its water, adds Steve Mumme. "I do think the price of water will start skyrocketing in urban areas, and that will put pressure on agriculture because farmers will eventually end up wanting to sell some of their water rights."

The specter of a mostly barren Owens Valley as the veritable hose that watered Los Angeles and its environs hangs over the heads of politicians, water administrators, water gods, and more. However, there's more to the story than meets the eye, says water wholesaler Kightlinger, whose organization has developed its own approach to purchasing water rights from agricultural interests.

"There's often a misperception of Owens Valley," says Kightlinger. "Obviously interbasin water transfers today are done much differently than they were in the early 1900s. Then, the city of Los Angeles went out and acquired tracts of land to acquire water rights, paid all the farmers for the land, bought them out, and developed that system. That approach worked a century ago."

Kightlinger and the Metropolitan Water District of Southern California have taken a different approach. "We've put together a program that seems to work very well with farmers in the Palo Verde Irrigation District," he says. "We approached the farmers and said we'd like to buy a thirty-five-year lease on their land; pay them for it; they will continue to farm it with certain restrictions; and every two to three out of ten years, we can tell them to fallow a third of that land. So effectively, we bought up the rights to fallow a third of the valley at any given time only so often. (For example, the program will fallow nearly 26,000 acres August 1, 2010, through July 31, 2012.)[5] "We pay them the profit they would have made had they farmed the land. We also paid them for the lease. The farmers like it because in an uncertain economic climate, we offered them a pretty steady income flow. They also liked it because we would not entirely fallow everything. We fallow a rotating third, which would rest the land and also provide water supply to us when we need it in dry years. So that was a more nuanced, sophisticated deal where you transfer agriculture water to urban [use] while preserving farming in the community," Kightlinger adds.

More such programs are being developed today, Kightlinger says, adding that he doubts water providers will simply buy out an entire valley, fallow the land, and walk away from it as they did a century ago. "Agriculture communities are eager and interested in today's programs because it's truly a win-win."

WATER TALES

 Along the Oregon-California border in the Klamath River Basin, water wars have brewed for years. The fight pits, among other things, the fish and the Endangered Species Act versus farmers who need water for their livestock and crops. Adding to the fray is another stakeholder, Portland, Oregon-based PacifiCorp, which operates a string of hydroelectric power plants on the river.

In 2001, the U.S. Bureau of Reclamation cut off most of the water supply to the Klamath Reclamation Project in accordance with requirements to protect fish habitat as part of the Endangered Species Act. That left farmers, who had for decades counted on the water, high and dry.

Years of dispute have followed. In 2001, water "mysteriously" was diverted to the "closed" irrigation canals. Meanwhile discussions and litigation continued. Early in 2010 with great fanfare, a "final" agreement was announced to settle the dispute and enable removal of four dams on the river. But not all parties in the fray were involved, so it's not over by a long stretch. Today reclamation of the Klamath Basin continues (in January 2011, the EPA approved California's formal water-quality improvement plan for the Klamath River), as do water allocations for irrigation in accordance with the Klamath Basin Restoration Agreement.

Majority rules. Different opinions and agendas aside, events, economics, and simple majority rule may force farmers eventually to re-evaluate how they use water, adds hydrometeorologist Frank Richards, who studies how the water on the ground affects the weather. "Decisions are made by the majority, and farmers are not a majority. They're always going to be at a disadvantage when decisions are made solely by a majority, especially if the decision process isn't informed and doesn't fully consider impacts on all the stakeholders," says Richards.

In the Southeast, for example, will Atlanta or rural areas get the water? In California, will it be the cities or the rural agricultural areas? asks Richards. "When there's a water shortage, the greater number of people in urban areas versus rural, agriculture ones, may well outweigh and outvote, and out-influence, the decision makers," says Richards.

Florida is another populous state straining its water supplies. One option is to convert land use from agricultural to residential, says water negotiator Mumme. Florida's agriculture is highly developed, with lots of water committed to it. "But as its population grows, it displaces that agriculture. Yet you still have water allocated and part of the concept is to convert that water to public supply use." Given the many demands on water, Mumme says it's "absolutely inescapable" that we learn to use it more frugally. That's slow to happen because the economics of consumption and development strongly favor past practices, and regulating water resources is not a popular thing, he adds.

BOTTLED DETAILS

Beyond agricultural and municipal water sources and usage, bottled water—consumed by billions of gallons a year (see Table 7.1)—has its own issues and controversies. They range from cost to content—toxins included—to costly containers. There's even a new kid on the block, Boxed Water Is Better, a Grand Rapids, Michigan-based company that peddles its product in old-fashioned containers in new-fangled ways. The company claims three-fourths of its box container is made from renewable resources; the boxes are shipped flat to regional fillers, and if that's not environmentally friendly enough, the company gives 20 percent of its profits to water relief and reforestation foundations. (Check them out at http://www.boxedwaterisbetter.com.)

The Real Cost

Unlike its tap-water cousin, bottled water is anything but cheap. Even the most fanatic aficionado of bottled water is quick to admit that. The lowest-price generic brand likely costs about $1 or so per bottle, while some of the high-end, *haute cuisine* varieties can set you back considerably more than that. To get a figurative taste of some of the newest waters out there, check out industry

TABLE 7.1   U.S. Bottled Water Market

	Volume and Producer Revenues 2000–2009			
Year	Millions of Gallons	Annual % Change	Millions of Dollars	Annual % Change
2000	4,725.1	—	$6,113.0	—
2001	5,185.3	9.7%	$6,880.6	12.6%
2002	5,795.7	11.8%	$7,901.4	14.8%
2003	6,269.8	8.2%	$8,526.4	7.9%
2004	6,806.7	8.6%	$9,169.5	7.5%
2005	7,538.9	10.8%	$10,007.4	9.1%
2006	8,253.5	9.5%	$10,857.8	8.5%
2007	8,757.4	6.1%	$11,551.5	6.4%
2008	8,665.6	−1.0%	$11,178.5	−3.2%
2009	8,454.0	−2.5%	$10,595.0	−5.2%

Source: Beverage Marketing Corporation.

publication *Beverage World* (http://www.beverageworld.com/content/category/31). Bottled Water of the World (http://finewaters.com) is another place to check out the plethora of bottled waters.

Whatever your preferences, the cost of bottled water becomes staggering when you factor in packaging, production, and the water required for those processes. The Environmental Law Foundation, an Oakland, California-based organization committed to improving environmental quality (http://envirolaw.org), found in a 2003 study that one bottle of water costs from 240 to 10,000 times more than tap water, depending on the origin of that bottled water. That group also reported that 10,000 gallons of tap water could be bought for the amount that one bottle of Evian water costs to produce and purchase.

Environmental Law Foundation numbers aside, realistically it takes water to bottle water. The resource is needed to produce the power to produce the petroleum to produce the plastic, to package the product, to ship the product, and so on.

However, to be fair, the International Bottled Water Association (IBWA) (http://www.bottledwater.org/default.htm) is quick to point out that bottled water containers are "fully recyclable." The industry group also reminds consumers that despite the reports by various groups on the pollution dangers of tap water, bottled water remains safe. After release of the Associated Press report of trace pharmaceuticals in tap water (discussed in Chapter 4), IBWA issued a statement reminding consumers of the safety of bottled water:

> The International Bottled Water Association (IBWA) would like to remind consumers that bottled water is not simply tap water in a bottle and that the safety and quality of bottled water produced in accordance with U.S. Food and Drug Administration (FDA) standards do not pose a health risk due to pharmaceuticals or other substances. Bottled water is comprehensively regulated as a packaged food product by FDA. Bottled water companies use a multibarrier approach to bottled water safety, which includes source protection, source monitoring, reverse osmosis, distillation, filtration and other purification techniques, ozonation, or ultraviolet (UV) light. The combination of FDA and state regulations, along with a multibarrier approach and other protective measures, means that consumers can remain confident in choosing bottled water.[6]

PRIVATIZING WATER

Is water a commodity to be bought and sold, or does everyone have the right to have access to water? That's the question that pervades most, if not all, of the battles over water. It pits East versus West, environmentalist versus rancher, farmer versus farmer, legislator versus legislator, neighbor versus neighbor, and so on. The answer isn't simple, given the different stakeholders. The path to consensus is even more complicated, confusing, and convoluted.

Underlying the arguments for and against water as personal property is the fact that water is by its nature a shared resource. At least, that's how Villanova University water law expert Dellapenna sees and teaches it. "The water I use today, you use tomorrow or vice versa, depending on who is upstream and who is downstream. You cannot freely deal with your water and ignore its effects on me," says Dellapenna. Regulating interactions among water users is the basic problem in water law, he adds. "If I'm upstream from you and free to sell my water to whomever, and damn the people downriver, an awful lot of people are going to find they don't have any water. That's problematic, and part of the dispute between proponents of prior appropriation and other approaches to water."

Of course, for every action, there is an equal and opposite reaction, and when it comes to water, there's always another side to the equation.

Western Commodity

Individual rights to control, buy, and sell water are a phenomenon of the U.S. West, where prior appropriation is king. Water is seen as a wealth builder, as part of a family's inheritance, and as a commodity on the open market. Not everyone agrees with the concepts of water as a commodity, however, and it's an ongoing battle that's not likely to end soon.

Property or not? Strict property ownership of water isn't what prior appropriation is all about, says Idaho's Clive Strong. He has spent 25 years dealing with water issues in that state, and says that water, unlike other goods and services, is owned by states in their sovereign capacity. "Unlike widgets, water can never truly be owned by any individual. Rather, it is the right to use water that forms the property right, which is granted by states," Strong adds.

"There's a misconception about the prior appropriation doctrine," says Strong. "People tend to think of it as simply *first in time, first in right*, and that's the end of the story. But the prior appropriation doctrine is multifaceted. First in time, first in right is an important element of the doctrine. Of equal importance is the issue of *beneficial use*. You must be able to demonstrate that the way you use the water is a beneficial use of water.

"In Idaho," says Dave Tuthill, "water is considered the property of the state. Individuals have the right to use water, and that right is limited to the extent of their beneficial use. For example, a farmer might irrigate fifty acres of ground and might have a right to what we call fifty 'miner's inches,' which is one cubic foot per second of water. But that water right is only for irrigation of those fifty acres. The water user cannot change the location, the point of diversion, or the place of use, or any element, without going through state approvals."

One of the classic cases in Idaho dates back to the early 1900s. A farmer was using a water wheel to lift water out of the Snake River Canyon to irrigate about 300 acres. Some junior water rights appropriators decided to build a dam that took away the current of the river. The water-wheel user sued, claiming he was first in time, first in right, and the junior users should be curtailed. The case went to the U.S. Supreme Court, which ruled against the water-wheel user, saying that even though he was first in time, first in right, he didn't have the right to monopolize use of the resource. The water-wheel user was not allowed to defeat the junior appropriators' ability to irrigate hundreds of thousands of acres.

"So," adds Strong, "beneficial use is an important element of prior appropriation, and people often forget that."

Dellapenna disagrees. "The Idaho system works only because the prior appropriation system is simply not enforced there. . . . Anyone who says something different is just not correct. That is part of the hype of pushing [open] markets for water."

One reason so much alfalfa is grown in the West, Dellapenna says, is that that's the use for which water was first appropriated. He points out that if the law is being enforced, someone who owns an appropriative right cannot change the time, place, or manner of use if it would have any adverse effect on other water rights holders, even junior water rights holders.

Tuthill calls that argument "simplistic," and says it has been rejected by the Idaho Supreme Court. "Some would like to reduce the argument to first in time, first in right, but our Supreme Court has appropriately said it's more complex than that," he says. Nonetheless, "some say we're not administering the water properly."

So much for arriving at an agreement!

Capitalizing on water rights: The Pickens equation. As we've mentioned, Texas oilman T. Boone Pickens and his Mesa Water have big plans to sell billions of gallons of water in the Texas Panhandle. It's a Texas-sized approach to free-market economics, and naturally, it has its detractors.

"Some of Pickens' claims about the right to sell water are just appalling," says Dellapenna, a proponent of regulated riparianism. "His idea is that he owns the groundwater and can do anything he wants with *his* groundwater. He can pump as much as he wants, and if it dries up his neighbors' wells, too bad for his neighbors. There actually is some support for that in Texas law, but that's a law that cries out to be changed."

Pickens likely understands the ramifications of such an approach to water, says Dellapenna. However, he adds, beyond the West, that's not the case, as illustrated by some farmers in Arkansas and Georgia who have notions of selling their water rights. "They don't seem to grasp it all. They say, 'It's my land, my water; I can do whatever I want.' But they don't think it through. Whether water is a property right or not—that's another whole layer of argument—at the very least, they need to realize their neighbor can do the same thing to them."

Lone Star Sierra Club's Ken Kramer doesn't cotton to Pickens's plans, either. "I don't think it is a viable plan economically in terms of some of Pickens's potential customers, like the Dallas–Fort Worth area," says Kramer. "Water managers from various utilities have said privately the cost of the water Pickens would provide is so far above what they would pay for water from other sources that it's just not economically feasible.

"The larger . . . picture is that it's yet another example of an effort to . . . deplete a groundwater resource," Kramer adds. "The Ogallala aquifer is being massively overpumped in Texas. Four million to 5 million acre-feet of groundwater are being pumped out of the Ogallala each year, and the recharge to the aquifer is only at best 400,000 to 500,000 acre-feet a year.

"As the water levels decline, the energy costs to pump that water will make it prohibitive for many agriculture enterprises to continue pumping," Kramer adds.

Pickens's basic argument is that if he doesn't pump out the water, someone else will, Kramer says. And Pickens is correct, given current groundwater management in Texas, which is based largely on the right to capture—if you own land, you can pump out what's under it, says Kramer. "Nevertheless, from an environmental standpoint of pushing for sustainable groundwater resources, that's not the appropriate approach."

Whether an individual does or doesn't like Pickens's plans to sell water, the basic concept of the haves selling water rights to the have-nots is here to stay, says Tuthill. Projects such as Pickens's are symptoms of the lack of water in some locations and the ability to move water from one location to another when there are economic incentives. "I think similar projects will increase over the years as we continue to put demands on water supplies that are, in many cases, already fully appropriated," Tuthill adds.

In a state like Idaho, however, drawing down an aquifer to the point that it "changes the character of the economic capabilities of a basin" is prohibited by law, says Tuthill. "Drying up an Owens Valley, as happened in California, is not allowed by statutes. It's an expression of a limitation on a transbasin transfer."

Capitalizing on Water Rights II. Another company with water to sell—"nearly two billion gallons filtered through rock into this quarry and protected by more than 400 surrounding acres" in upstate South Carolina—is Carroll Properties Corporation. The company is helmed by noted social entrepreneur Elizabeth C. Belenchia, who has been involved with United Nations and environment and sustainability projects around the world. The price for the property, water included, is $29.5 million, though the property owners are considering other options, including, says Belenchia, sale of the water separately to other entities. (Check it out at http://clearwaterofcherokee.com/.)

Stay tuned for more water-conscious business ventures in the future. As urban sustainability expert Williams suggested in Chapter 3, water is the limit to growth.

Et Cetera

As the price of water increases, experts expect to see new and unusual schemes for buying, selling, and transporting it—selling water to China, towing icebergs from Alaska, and piping water great distances from the haves to the have-nots. Mumme recalls one proposal for the city of San Diego that involved floating trains of water bladders (giant rubber-type balloons) from Alaska down the Pacific Coast and then anchoring them off San Diego. An engineering firm designed an entire fleet to do the deed, says Mumme. "What seemed like preposterous proposals will resurface now because of the issues related to climate change and water scarcity," he says, "particularly some focused . . . on urban needs for water."

But, adds Jeffrey Kightlinger, "I don't think we're looking at icebergs at the moment. An engineer once told me you could probably get the thing moving into the right place, but it would be very tough to stop it."

One geoscientist at the University of Nevada–Las Vegas proposed turning formerly mothballed warships into floating desalination plants. There's a precedent for this: National Guard mobile water-purifying systems have been used successfully in Iraq as well as in states like Iowa in times of flooding, says Iowa rural water expert Huff. Whether the United States is ready for such unusual water-related actions is open to discussion, though today there's plenty of talk about these floating desalination options. Another new entrant with grandiose plans to sell water, this time from Alaska, is S2C Global Systems (OTCBB: STWG), a San Antonio, Texas–based company with subsidiaries in Alaska, British Columbia, and Nevada. The company touts itself as focusing on the export of "billions of gallons of water globally from the watersheds of Baranoff Island, Alaska." In July 2010, the company announced it was within months of beginning distribution from a "World Water Hub" on the west coast of India.[7] Whether that's a done deal is yet to be determined. Remember the big plans of the Canadian firm that wanted to export Lake Superior water to China. Public outcry shot that down back in 1998.

Native Americans in the Water Equation

Other possible players in the future of water as a commodity are Native American tribes in the West. Many already have profited from their federally reserved senior rights to the nation's water. Time and again, courts have upheld their water rights as senior to, or taking precedence over, other state water rights. Recall the Wind River Reservation's court-approved claim for hundreds of thousands of acre-feet of water (mentioned in Chapter 5). Other Indian reservations potentially hold tremendous power over water, too. And power means rights to water and money.

The Native American tribes in Idaho, for example, already are active in water marketing, says Tuthill. Most of these tribes have negotiated their water rights by going through the legal process to clarify the annual volume of water they're entitled to under their federally reserved water rights, and are an important factor in that state's water marketing, he adds.

A Wealth Builder

Further entrenching water as a commodity and widening the chasm between haves and have-nots, and between West and East, is the fact that many water rights owners in the West see their water as a source of real, lasting family wealth.

John Gandomcar, the modest Colorado developer and horse farm rancher, is typical of many water rights owners in the West. He considers his water rights an integral part of his wealth and a future inheritance for his children. "I'm one of only five owners of water rights from a sizable reservoir in the mountains west of Denver. One of those other rights holders is the Denver Water Board, which serves metropolitan Denver. Those rights are important and valuable from an ownership point of view, and I plan to pass them on to my daughters," says Gandomcar.

Idaho water rights owner Newcomb thinks much the same about his water. After all, he says, water helped him, his father, and thousands of others across the western United States develop their lands and build wealth. The proof is in the rich farmland and ranchland today in states like California, Idaho, Nevada, and Arizona. The dams built by the U.S. Bureau of Reclamation, and subsequent irrigation systems, many built by farmers and ranchers, helped transform the high-elevation arid country into productive land, says Newcomb.

"We have about three and a half million irrigated acres of land in Idaho," says Tuthill. "The value of each of those irrigated acres is maybe $5,000. Without water, the value goes to almost zero."

"Any time you own water, it's a wealth builder," adds Newcomb. "I am a strong believer that land—as long as you're not land poor—will always have value because there's only so much of it. Water is much the same, and water may have even more value because it is diminishing in quantity. It's basically pure supply and demand. As time goes on and there is less water, it becomes more valuable.

"I know of one farmer/rancher about seventy miles from here who sold all his groundwater rights to SEMPRA, a California coal-fired generation outfit. They had planned to build a plant in the Jerome, Idaho, area," says Newcomb. The plant construction plans fell through, but Newcomb's fellow farmer still walked away with "a significant amount" of cash.

WATER TALES

Bruce Newcomb recalls the early years when he and his family turned arid land in the West into rich ranchlands with the help of water, and how they built wealth in the process.

"Basically you took a piece of ground that was fairly cheap as dry land grazing. That's what my father did when he first started and what I did when I first started. I paid the state of Idaho $50 an acre for some of my land, then I had to drill wells, put in the sprinkler pipe and mainline. But when I was done, I had ground worth $300 to $350 an acre. And I could irrigate it, raise potatoes and wheat. . . . It was basically taking something of no value and making something of value.

"At the time, there also were all kinds of state and federal incentives to transform the land. Utility power companies even offered incentives if you would just drill water wells. Some even paid for your pumps to pump out the water in exchange for becoming your power provider."

Future Change Possible

However, even Newcomb admits that in the future he may not have as much control over his water. That could be the case if, for example, public policy dictated that because water is a public resource and people can't live without it, it must be regulated so no one can tie it all up, with potentially fatal results. Newcomb reconsiders, though, and adds, "I think we basically have that now." Nearly every state has a department of water resources or something similar that prevents people from buying up water rights and not allowing anyone to use them.

"The problem with simply making water a commodity," the Sierra Club's Kramer adds, "is that you lose the perspective that water is a basic resource everyone must have. It's not like other products that may or may not be essential to human life or well-being. As human beings, we must recognize that water is an essential element of our lives and not discretionary."

Water Banks

Western states that follow some form of prior appropriation doctrine generally also have adopted the concept of water banks to solve increasing market demands and to allow an exchange of water within the constraints of existing rights. It's a market-based approach that allows those who have water rights and don't need them to "bank" them so they're available to others for a price, and generally for a set time.

Water banks are an alternative to the use-it-or-lose-it aspect of the prior appropriation doctrine, and are a device sanctioned by some states, says hydrometeorologist Richards. They're also an example of trying to manage water better during a time of tight water and within the constraints of our current legal system, he adds. He doesn't think they go far enough, however. Eventually the rules themselves may have to change. "What will happen is what's happening in California already with its water bank: When things get bad enough, the governmental organizations will come in and work with the stakeholders," says Richards.

A transaction with a water bank isn't necessarily as quick as an ordinary transaction on a commercial commodities exchange, says Idaho's Tuthill. In Idaho, for example, the water bank is run by the state. Water exchanges can be for a set period of time, or rights can be bought and sold, which requires an in-depth water-transfer process. "We have to ensure that when water is changed in some way the water right is not enlarged, no one is injured, and it's in the public interest."

Dellapenna, on the other had, calls water banks state management masquerading as a market. "It's hyped that markets are the solution. The examples people point to as proof markets work aren't really water markets. For example, Arizona's water bank,

California's water bank, the so-called sale of water from the Imperial Valley Irrigation District to the city of San Diego in California, those weren't sales. You put a gun to someone's head and you say, 'sell,' that's not a sale; that's not a market," he adds.

Water banks are not really markets because the state decides who must sell and who gets to buy, and the state sets the price by administrative decision. Water banks *are* profitable, Dellapenna says, and that's one of the reasons there is a great deal of hype pushing them as markets. However, Idaho's Strong says, "Water banks will always be subject to some degree of state regulatory control because the states actually own the water, and individuals only own the right to use the water."

WATER—A COMMODITY ON THE MARKET?

Beyond water banks and the buying and selling of water rights in the West, water is in demand by ordinary investors, too. Almost every natural resource economist who works with water will agree that water is undervalued in the market, says Colorado State University's Mumme. In investment terms, that means room for growth, and where there's growth, there's generally the potential to cash in.

No Ordinary Commodity

Water, however, is not like other commodities. "Water is not a commodity you can buy or sell like gold or tin or lumber or pork bellies," says Russell W. Bauman, senior vice president and branch manager of David A. Noyes & Company, one of the largest full-service brokerage firms in Indianapolis, Indiana. "But there are areas where water is in short supply and people are looking for pure plays in water (investments in which the majority of company revenues come directly from water)," says Bauman. "People seek water as an investment because they read about shortages in many areas; a shortage always means demand, and demand greater than supply means the price goes up."

Nonetheless, even with all the warnings about water and water shortages, Bauman says he's had very little demand for water equity positions.

Commodity or not, and water rights and boards aside, Bauman says water isn't likely to end up beside pork bellies and tin on the Chicago commodities exchange any time soon. "I don't know that that could happen because water is readily available to anybody for free. Just put a great big funnel in your backyard and you will have all the water you want. The water that falls in your yard is your water—unlike other commodities, such as gold. [That is, unless you live in a state where the water from the sky doesn't belong to you!] You can't just go out and dig for gold, because there's no way you can assay it and refine it and so forth. But water is readily available," he adds.

The business of water expert Maxwell has a slightly different take. "The one thing which I think has become clearer during this recent recession is the additional financial and investment attributes of water and water stocks. Water-related investments behave differently in an economic downturn. They tend to be much more buffered from external economic circumstances than most other types of investments simply because our water usage doesn't really depend much on economic circumstances.

"In that regard, I believe that water will start to be viewed as almost a separate type of investment class in the future—whether as physical water, water utility stocks, or the stocks of companies that provide goods and services for the provision of water. In the future, water may come to be viewed more as a commodity or as almost a 'store of value' as we view gold or silver today. I don't think it's too much of a stretch to imagine that 50 or 100 years from now we may see a reversal of some of the migration trends that have shaped this country in the past 50 or 100 years. People will start to move away from the arid Southwest, away from the boom towns like Las Vegas and Phoenix, and back to areas like Cleveland (Ohio) or Buffalo (New York) where water is more abundant," Maxwell adds.

Stocks, Mutual Funds, and More

Various stocks, mutual funds, and unit investment trusts count water-related investments as part of their portfolios. That investment may relate to water utilities, infrastructure, conveyance, treatment, monitoring, or legal compliance in the United States or around the globe. Several global investment indexes follow water-related companies and industries, including:

- Standard & Poor's Global Water Index (http://www .standardandpoors.com/indices/sp-global-water-index/en/ us/?indexId=SPGBTHSUW-USDW—P-RGL—)
- Palisades Water Indexes (http://palisadesindexes.com/ waterindexes.html)
- International Securities Exchange (ISE) Water Index (http:// ise.com/WebForm/options_product_indexDetails.aspx?categ oryid=234&header0=true&menu2=true&link2=true&symbol= HHO)
- Janney Water Indexes (http://janney.snetglobalindexes.com/ industry.php)
- Global Water Intelligence Water Index (http://www .globalwaterintel.com/water-index/)

Here's a sampling of some big publicly traded water-related companies.

- Veolia Water (NYSE, VE): A leading water and wastewater ser-vices provider, Veolia Water, which serves approximately 163 million people in 66 countries worldwide, is part of France-based Veolia Environment (http://www.veoliaenvironnement .com; http://www.veoliawaterna.com).
- American Water (NYSE, AWK): This New Jersey–based firm is the largest publicly owned operator of water and waste-water facilities in the United States, serving approximately 16 million people in 35 states and two Canadian provinces (http://amwater.com).
- Aqua America (WTR, NYSE): Based in the United States, this water and wastewater utility serves nearly 3 million residents in 14 states, including Pennsylvania, Ohio, North Carolina, Illinois, Missouri, Texas, New Jersey, New York, Indiana, Florida, Virginia, Maine, Missouri, and South Carolina (http://www.aquaamerica.com).

Hundreds of mutual funds had water-related holdings (water utilities, transportation of water, etc.) in their portfolios as of September 30, 2010, according to analysis by Morningstar, Inc. But, of the more than 675 stocks tracked by Morningstar for this book, only 5 had more than 10 percent of their portfolio's holdings in water-related stocks. They include (© 2010 Morningstar, Inc.):

- Calvert Global Water A (CFWAX): 23.55 percent of the fund's assets are in water utility stocks; market value of fund is $7.80 million.
- Allianz RCM Global Water A (AWTAX): 22.38 percent of fund's assets are in water utility stocks; market value of fund is $11.21 million.
- Kinetics Water Infrastructure Adv A (KWIAX): 20.09 percent of fund's assets in water utility stocks; market value of fund is $5.45 million.
- PFW Water C (PFWCX): 16.24 percent of fund's assets are in water utility stocks; market value of fund is $3.22 million.
- New Alternatives (NALFX): 11.08 percent of fund is in water utilities; market value of fund is $27 million.

Certain unit investment trusts provide another approach to water-oriented investing. These are unmanaged portfolios (unlike mutual funds which are actively managed) of securities held for a set period of time, which investors then purchase by the "unit," and they're growing in popularity. For example, Guggenheim, an asset management firm, already has sold nearly $542 million of its Global Water Equities Portfolio (CGWERX, NASDAQ), says Bauman. The trust's portfolio is made up of 40 securities of companies that focus on various aspects of the global water business.

WATER REALITIES

- As supplies have become scarcer and infrastructure needs have grown, water has become the hot commodity of the twenty-first century.

- The fundamental (and controversial) question of whether water is a God-given right or a commodity to be bought and sold isn't likely to be answered any time soon.
- Meanwhile, water rates will continue to rise, especially when staggering costs to replace antiquated infrastructure are factored into the equation.
- What Americans pay for water doesn't always reflect its value or even the real cost of delivering it.
- Water is firmly entrenched as a wealth builder and provider in the West, where prior appropriation is the water law.
- Individuals familiar with water laws and flush with water capitalize on others' need for the resource through private and public exchanges (for example, water banks).
- Investors look to stocks in water-related companies and industries, and funds holding such stocks, because shortage combined with demand can equal profit potential.

NOTES

1. Beverage Marketing Corporation, press release, "Bottled Water Confronts Persistent Challenges . . . Report from Beverage Marketing Corporation Shows" (July 2010), http://beveragemarketing.com/?section=pressreleases.
2. George A. Raftelis, "Water and Wastewater Finance and Pricing: A Comprehensive Guide," Raftelis Financial Consultants, http://raftelis.com/ratessurvey.html; American Water Works Association and Raftelis Financial Consultants, "2010 Water and Wastewater Rate Survey (Bi-annual)," http://www.awwa.org/Resources/Surveys.cfm?ItemNumber=39303&navItemNumber=39456.
3. U.S. House of Representatives Testimony, January 25, 2010, "Testimony before the Subcommittee on Water and Power of the Committee on Natural Resources of the U.S. House of Representatives, Testimony of Dr. Peter H. Gleick . . . ," Summary, http://resourcescommittee.house.gov/images/Documents/20100125/testimony_gleick.pdf.
4. Pacific Institute, "Saving a Million Acre-Feet of Water through Conservation and Efficiency," press release, September, 8, 2010, http://pacinst.org/reports/next_million_acre_feet/next_million_acre_feet_release.pdf.
5. Metropolitan Water District of Southern California, July 27, 2009, "Metropolitan's Fallowing Call Starting August 1, 2010," http://data.pvid.org/MWDDocs/LinkClick.pdf.

6. International Bottled Water Association, position statement, "Bottled Water Is a Safe, Healthy Packaged Beverage Choice," March 11, 2008, http://www.bottledwater.org/public/2008_releases/2008-03-11_statement.htm.

7. S2C Global Systems, "S2C Global Announces India World Water Hub," July 7, 2010, press release, http://www.s2cglobal.com/2010/07/a2c-global-announces-india-world-water-hub/.

8

CAN OUR WATER BE SAVED?

Human history becomes more and more a race between education and catastrophe.

—H. G. Wells
The Outline of History (1920)

Now that you've heard the doom and gloom about our nation's water supply, you're probably either scared that the United States is drying up, paranoid that the water from your tap is hopelessly polluted and could make you deathly ill, or a combination of both. Is there any hope for the future of water in the United States?

There are no easy answers. Environmentalists, big corporations, municipalities, metropolitan areas, farmers and ranchers with water and those without, researchers, states, courts, and the federal government all have different viewpoints on the future of America's most precious resource. Individuals are left in the crossfire. When it comes to water, consensus is a tough sell. After all, everyone has a vested interest—we all need water to survive. Yet, as the many tales of water success reflect, compromise and consensus are not unachievable.

Peter Gleick of the Pacific Institute stops short of calling the water situation across the United States a crisis. "We have serious water problems, but, unlike the global water problem, we have the money, the resources, the technology, and the institutions to manage these problems," he says. "I'm not saying we do manage them, but we could." At the moment, however, Gleick cites underinvestment, state-to-state conflicts, serious ecosystem degradation, and the risks of growing climate change as factors affecting water supply that need addressing.

"We are up against the limits of supply in the western United States," he adds. "We still think about water in the western United States the way we thought about it one hundred years ago, which is 'let's just find another aquifer to tap or another river to dam.' That kind of thinking will not get us out of our problems. Potentially, we could have a very serious water crisis if we don't do things differently."

Let's look at some potential solutions and ways of doing things differently, along with the associated costs, conundrums, and confusion.

RECOGNIZING THAT THE PROBLEMS EXIST

The biggest obstacle to solving the nation's water problems is refusing to admit they exist. Even now in the face of serious concerns and in places like Southern California where they're gasping for water, many people won't admit to the problem, let alone change their water habits. It's soak the lawn, run the water, dump the waste as usual.

To be fair, many aspects of the national water crisis are invisible. When you turn the faucet, water comes out. No one can see arsenic, atrazine, or antibiotics in the water, or leaking sewers leaching away good water with the bad. People often fail to realize, until it's too late, that children are being poisoned by lead in old pipes, or that a well has run dry, or that out-of-sight underground pipes are crumbling. Neither do they see the standing water in the corner parking lot's expanse of asphalt as anything more than a nuisance and habitat for mosquitoes. (More important, it's also a red flag that underground water supplies aren't being replenished.) Lush green lawns and growing water-intense crops in arid climates fed with millions of gallons of potable water also remain standard operating procedure for most as opposed to the exceptions they should be. Until now water shortage was little more than a tug at the edge of our collective consciousness—certainly not a real problem. Slowly, however, the reality that something is happening has begun to creep in. Nearly every day we read about another water main break or sinkhole somewhere. In many places temperatures may seem a little warmer or colder, precipitation a little more or a little less, or the weather may just feel "different" from that of years past. Chances are we've all heard people talk about "the huge snows we used to have" or complain that "the ground is so dry, or "this is the worst storm I can remember."

Drought has affected vast areas of the country at the same time that development has exploded. Crisis water shortages in cities like Atlanta, Georgia; Tampa, Florida; Richmond, Virginia; Denver, Colorado; and Pittsburgh, Pennsylvania have filled the headlines. Suddenly, the crisis hits home—*it can and is* happening here!

Today water is an issue for everyone. It's no longer the sole concern of the poor in Central and South America, Asia, or Africa, and it's not just lapping at the back door of the United States. It's banging on our front door—and, in some places, barging into the living room, kitchen, and bathroom. It dries up farm fields and cities, and the lack of water or attention to it often results in pollution of lakes and streams. It drains our pocketbooks and threatens to reach the point of no return.

Plentiful rains no longer are enough to stem the tide of our water problems either. In early 2009, for example, it seemed that heavy rain ended Atlanta's years-long physical drought. But not so fast. "The one thing I can definitely forecast is drought will return—whether it's in one year, two years or more," David Stooksbury, Georgia state climatologist and professor of engineering and atmospheric sciences at the University of Georgia in Athens, said at the time. "It's imperative for people to know drought is a normal part of the climate system."

Stooksbury's words proved all too true when in mid-August 2010, he announced a return to drought conditions in the north-central, west-central, and southwest parts of Georgia. In November, Stooksbury followed up with more dismal news: Drought now covered most of Georgia, with little relief expected in the near future. "The fall and spring climate outlooks do not hold much promise for drought recovery. Climatic conditions are expected to be warmer than normal and drier than normal through the spring."[1] Couple that with Atlanta's Lake Lanier woes we talked about earlier, and it becomes apparent that water-wise, the going is rough for this gateway to the South. Atlanta even has a web site dealing with its water issues, http://atlantawatershed.org/drought.

"We cannot continue to consume water at the rate we have been consuming it," says Las Vegas water czar Pat Mulroy. "We have to stop taking water for granted."

With Mulroy's warning in mind, let's consider possible ways to ease the water crisis.

NO SIMPLE SOLUTIONS

Until now, the solution to some of the nation's water woes has all too often been ad hoc adjustments, says hydrometeorologist Frank Richards. "We need to bring to bear objective scientific information so society can develop a more systematic approach."

Water and the multifaceted problems associated with it require multifaceted solutions. It's not enough that the United Nations declare World Water Day or World Water Week; that the Great Lakes states and Canadian provinces sign a water compact; that environmental groups, municipalities, utilities, or watershed and river basin agencies take up a cause; that individuals crusade for their cause; or that states and public or private groups square off in court.

"There is no magic wand, no flip of the switch that is going to suddenly eliminate water scarcity," said Jacques Diouf, director-general of the United Nations Food and Agriculture Organization (FAO), in March 2007 in conjunction with World Water Day that year. "But there are concrete ways to turn the tide against water shortages."

Issues that must be addressed to cope with water shortages range from protecting the environment and global warming to fair pricing of water services and equitable distribution of water for irrigation, industry, and household use, Diouf says. Agriculture must take the lead in coping with water scarcity by finding more effective ways to conserve rain-fed moisture and irrigate farmlands, he adds.[2]

The solutions aren't instantaneous, and many aren't easy. But solutions can be found. The ideas, like the problems, seem endless, and often contradictory and controversial. Some approaches—like water conservation, reuse, increased efficiency, cleanup, education, and more—are realistic and achievable. Other ideas—like capturing icebergs or floating giant water balloons to points elsewhere—are perhaps less so. With all the proposals and suggestions, though, the theme is change. Status quo when it comes to water in the twenty-first century is a concept of the past. As we've discussed throughout *Aqua Shock*, some forward-thinking metropolitan areas, municipal water districts, states, public and private organizations, and individuals already have embraced change and are spearheading or have implemented various water initiatives with success to their credit.

"Water will never be an issue that will be solved once and for all," says Iowa's Greg Huff. "There will always be challenges, both on the local and regional levels. We must continue to seek ways to address those challenges and find and implement solutions that are appropriate and effective for the problems facing each particular area and situation."

Options, Opinions, and Answers

There are few silver bullets to solve the nation's water crisis, says Sandia's Mike Hightower, who also travels the country talking to and working with experts on how to make the most of the resource.

Limits to new sources. "Our efficiencies and new sources of water won't be from new rivers, new aquifers, or any new freshwater sources," says Hightower. "Instead, they will be from better management, better coordination, more efficient use of the water we already have, and additional use of nontraditional or alternative water resources like reusing wastewater or desalination. These ideas will become the mantra for water use and water development in the United States in the next 50 years. We don't have any choice."

Some states, like New Mexico and Colorado, already have embraced the mantra. New Mexico, for example, has a 50-year state water plan. Looking to the future, that state's engineer has said that if power plants are built to supply electricity to California and New Mexico and water is used to produce the power, then

New Mexico should get water from California to support producing that power, adds Hightower. "People are beginning to understand that exporting electrons to another state is the same as exporting water."

Needs versus Demands

A few thousand miles to the east in Massachusetts, Kerry Mackin expounds on how to ease the nation's water woes. She's head of the Ipswich River Watershed Association. Remember, the Ipswich is the river that periodically ran dry in sections during the summer, in part due to overpumping of groundwater. "The solution is to look at the optimal way to meet our real water needs as opposed to our demands," says Mackin.

Her multipronged approach includes these suggestions:

- *Use water as efficiently as possible,* including lowering nonpriority uses like lawn watering so they don't supersede fisheries and other essential uses. This reduces energy demand, too.
- *Optimize water-supply management scenarios.* That could include a mix of groundwater and surface water, or using surface water supplies in the summer and groundwater supplies the rest of the year.
- *Offset or mitigate water demand.* These offsets should include programs to capture storm water and return it to replenish groundwater aquifers rather than allowing it to run off. Another offset could be to treat wastewater locally and return the cleaned, treated water into the ground to replenish aquifers.
- *Include environmental costs in the price of water.* Returning some of the costs of water extraction to restore damaged rivers will help reduce demand and provide economic incentives to balance the water budget.

"The key, in our basin at least, is to restore the groundwater hydrology by getting water back into the aquifers," says Mackin.

Aquifer replenishment has grown in popularity worldwide as one aspect of managing groundwater resources. In the United States, it's catching on, too, and has been implemented successfully in water districts across the country, including in Texas, New Mexico, Nevada, Florida, and California.

Time to Stockpile or Rethink

Despite the hype, the threats, the fights, the bickering, and the court cases, all-out water wars à la guns drawn at the OK Corral aren't likely. Neither is the necessity to buy a huge underground storage tank and fill it with water, reminiscent of the oil-hoarding energy crisis of the 1970s. "We must think more carefully about how we *need* to use this resource—not how we *want* to use this resource," Colorado State University's Steve Mumme says. Further echoing Mackin's admonitions, he adds, "Needs and wants are very closely connected when you get right down to it. People who want in-stream water flows so they can go kayaking or water rafting are dealing with an amenity use of water that gets all flummoxed up with farmers' desire to grow feed grains in, say, Greeley, Colorado. It's all the same water, but there is only so much, so we have to think about what our needs actually are.

"Water is a scarce resource, and we have to think about both pricing and regulating it more effectively. It may not make a lot of sense for urban areas to grow bluegrass lawns. Low-water xeriscaping—landscaping using drought-resistant plants—might work well instead," Mumme adds. "Equally, it may not make much sense for farmers in the Rocky Mountain West to grow sorghum and feed grains to fatten cattle, a very low-value use of water."

A shift away from tradition, however, isn't popular with agriculture groups because generations of families have had the freedom to use water as they wished, and they're accustomed to growing and selling things in a certain market. It's as much a lifestyle as an economy, says Mumme. "The question I have is, can we afford to sustain that lifestyle?"

Hydrometeorologist Richards agrees that ultimately societal change must occur to cope with water scarcity. "I don't think we can expect to address the problems by somehow modifying natural climate variations."

Instead, the solutions involve better management of water as the shared resource it is. That translates in some areas to growing crops that perhaps aren't as valuable, but don't require as much water, experts agree. How realistic such societal shifts are depends on whom you ask. Remember second-generation Utah farmer Jerald Anderson, who opposes the proposed removal of Snake Valley water by the Southern Nevada Water Authority? Anderson and many fellow Utah farmers fear the removal of water from their shared, cross-border aquifer will dry their fields and destroy their livelihoods. "Many people here who can survive in this lifestyle really don't have alternatives if their fields dry up and farming doesn't work," says Anderson.

Tennessee hydrologist William Waldrop, an expert in groundwater flows, also points to another, less widely recognized pollution threat from irrigation in the naturally low-water West. When an area like Southern California irrigates year after year, naturally occurring salts in the water accumulate and eventually will pollute the soil to the point that it may not be able to sustain agriculture because there isn't enough rain to wash the salts away.

These and other water issues will continue to worsen until we make some major changes in our use patterns, says Waldrop. "If you fly over Phoenix, Arizona—essentially a desert—it seems like everyone has a swimming pool. But if you live in the desert, I don't know that everyone can afford [water-wise] to have a pool. These kinds of quality-of-life issues have to be prioritized."

Planning Makes a Difference

El Paso, Texas, once was fast running out of water. In fact, in 1979 the Texas Water Development Board warned the city that its primary water supply, the Hueco Bolson aquifer, would run dry by 2020 if groundwater pumping continued at current levels.[3]

"That's not the case anymore," says El Paso's water administrator, Ed Archuleta. "We practice what I call total water management in El Paso. I believe that water is water. It may not be the quality you want, so you may have to treat it. It may be owned by somebody else, so you have to acquire it. It may not be where you want it, so you have to move it. But conservation is at the heart of everything we've done, to the extent that if we can conserve more water, that takes demand off the future water supply."

Based on their long-term planning, even with population growth, they don't have to do anything else for at least 30 to 40 years, says Archuleta. "When I came here almost 20 years ago, we relied mostly on groundwater, which was a declining resource. We devised a 50-year plan, and we followed it."

In 1990, El Paso residents used more than 200 gallons of water per person per day. Today, they use 133 gallons a person, despite an added 180,000 people in the area, says Archuleta. (Remember, water use is higher in more arid climates.) To achieve that water efficiency, the utility invested in elaborate groundwater models to provide a snapshot of water flows and aquifer levels, and then developed a multifaceted, *implementable* water-use plan that met the city's water needs. That plan included the following components:

- Setting rate structures to increase the cost for high water users
- Promoting water conservation through various incentive programs, including cash rebates to customers for installing central refrigeration units, low-flow showerheads and toilets, horizontal washing machines, and outdoor xeriscaping
- Increasing the use of water from the Rio Grande, as opposed to relying primarily on groundwater pumping
- Desalination of naturally occurring brackish groundwater
- Expanding the use of nonpotable, treated wastewater

"We also have quite a bit of reclaimed water that was part of our long-term strategy," says Archuleta. "This is not gray water, such as showerhead water and noncontaminated home water. This is wastewater, sewage that has been treated to a high standard for nonpotable uses."

That treated water is distributed to major users like golf courses, parks, cemeteries, car washes, construction projects, and apartment complexes, mostly for turf irrigation. Some goes to the power plant for industrial cooling. Using that treated water efficiently also reduces the demand on freshwater supplies, especially from large turf irrigators, Archuleta says.

Similar strategies will increasingly come into play for water reclamation and reuse in the western United States and in many parts of the Southeast and Midwest, adds Archuleta. More desalination plants are likely, too, and not only in Texas, California, or Florida. "I think you will see many more inland desalination applications like ours."

Diversify, Diversify, Diversify

Diversifying sources for water is important for cities, towns, and municipalities to consider when planning for growth, development, and need, says Archuleta. Don't put all your eggs in one basket or you could end up like Las Vegas, he warns. That city depends primarily on water from the Colorado River (and its Lake Mead). As mentioned, it supplies 90 percent of the city's water supply—and with lake levels way down (see Figure 8.1), what water there is already is appropriated. Combine those factors with the city's tremendous growth, and Las Vegas is scrambling to find water elsewhere to meet its needs.

Las Vegas' Pat Mulroy has worked zealously to diversify her authority's water resources despite drawing the ire of many in the process. She's also spearheaded massive conservation and construction initiatives to plan for catastrophic water shortages. "This nation is a voracious user of natural resources, and we have to knock it off," Mulroy says. "The days when we had a great expanse of country and few people and few food needs and sufficient agriculture property to meet those demands are over. Our population has grown, and it's going to grow by another 200 [million] to 300 million people. It's time to start thinking the impossible is possible," adds Mulroy. "Everyone assumed in the '90s, based on probability studies, that a drought of this magnitude would be impossible. So much for probabilities."

Water leaders, communities, states, and the nation must plan now for the worst-case scenario, she adds. "I'm not saying 'spend money.' But know what it's going to take to meet the challenges of whatever the *impossible* is; know what timelines are required to build that necessary facility; know what triggers and warning signs to watch for; and when those occur, step into action."

FIGURE 8.1 Low Water Levels in Lake Mead
Source: Las Vegas Valley Water Authority.

WATER TALES

 In late 2009, Pat Mulroy, who heads up the Southern Nevada Water Authority, talked about the harsh reality of the water crisis facing those who count on the Colorado River for water supplies. It's dominated by the specter of a dried-up Lake Mead with water levels too low to produce electricity or meet the ongoing water needs of her community. As she describes it,

Over the last nine years, we've had 66 percent of average spring runoff. That runoff is supposed to supply the lake with water. Four more years of this, and we lose our upper water intake into Lake Mead; six more years and we break water elevation 1,000 feet. We have two water intakes in Lake Mead. The lake is full at elevation 1,204 feet. Our oldest intake is at elevation 1,050. Our lower intake, installed in the mid-1990s, is at elevation 1,000 feet. We are currently in the process— at a cost of $1 billion—of installing a third intake that will go under Lake Mead and come back up at elevation 860 feet in the dead pool [the isolated water that's left when lake levels drop too low to permit water flow out of the lake], though it will only be operational at 1,000 feet. To put that in perspective, at water elevation 1,050, Hoover Dam stops generating electricity because [water is] not high enough to get into the turbines of the dam. At elevation 1,000, [there is] less than 5 million acre-feet left in a reservoir that holds 25 million acre-feet, and the annual draw on Lake Mead is 9.5 million acre-feet.

You can't build your way out of this unless you start thinking differently, and I think you'll see the panic button and the crisis button pushed in the Colorado River Basin the minute the first round of announced water shortages occurs at elevation 1,075.

We're at elevation 1,100 right now, and we figure that even with normal hydrology this year, we'll be down to 1,095. If we have a bad year, we'll drop even farther.

As of early September 2010, lake levels had dropped to 1,085 feet with the lake at just 37 percent of capacity. Following heavy rains, the level had climbed back to 1,087 by early January 2011, according to data from the U.S. Bureau of Reclamation (http://azgfd.gov/h_f/edits/lake_levels.shtm; http://www.usbr.gov/lc/region/g4000/hourly/rivops.html).

The theme for the future must be to accommodate change, whatever that change might be, says Richards. "Some of the things we're doing now must be different going forward given our water supply, climate change, and more," he says. Whether it's the Great Lakes compact, water law in the western United States, water law in general, or scientifically based water management strategies, it's "inevitable that we confront those issues."

CONSERVATION'S ROLE

J. C. Davis, a resident of Boulder City, Nevada, since 1994, is typical not only of the average Nevadan but the average American in that he hasn't really felt the impacts of drought or water shortages. Although he's not an environmentalist, he does like to cut down on unnecessary household chores and save money in the process. His solution: Ditch the water-guzzling lawn, and opt instead for native landscape.

"When I bought my current house in 2000, it was a turf landscape—okay, mostly weeds—and was sucking upward of 40,000 gallons of water a month," says Davis. "I converted the landscape to drought-tolerant, low-water-use xeriscape not because I'm a water guy, but because I hate mowing and edging when it's 118 degrees outside. I love the look, love the maintenance, and haven't used more than 6,000 gallons for watering the lawn in any month since," he adds.

In El Paso, Texas, efforts—including financial incentives—to promote water conservation have meant more than 3,000 homeowners have removed more than 9 million square feet of grass and replaced it with water-efficient landscapes.

For water savings, the U.S. Environmental Protection Agency's (EPA's) WaterSense program advises lovers of lush lawns to consider watering a little less. Test your lawn to determine whether it needs water. Step on a patch of grass; if it springs back, it doesn't need additional water.

Plenty of online sources can clue you in on what plants and grasses are native to your area, and optimal ways to make use of that vegetation. Here are a few:

- Native Plants Database of the Lady Bird Johnson Wildflower Center at the University of Texas at Austin (http://wild

flower.org/plants/): easily searchable database by state, habitat, lifespan, and more.

- U.S. Bureau of Land Management (BLM) Plant Conservation Alliance (http://www.nps.gov/plants/): Click on "Native Plant Landscaping Guides & Websites" for various options.
- Plant Conservation Alliance's wiki (http://www.plantconserva tionwiki.org/wiki/Native_plant_landscaping_information): state information and other useful links.

WATER TALES

The Southern Nevada Water Authority, a leader in consumer water conservation, urges its customers to save water and cash by replacing ordinary, water-devouring grass and shrub landscaping with water-smart trees, shrubs, and flowers. Every square foot of the converted landscaping saves 55 gallons of water a year.

As incentive, its Water Smart Landscapes program offers rebates to customers of $1.50 per square foot of grass removed and replaced with desert landscaping (low-water-use xeriscaping, for example) up to the first 5,000 square feet converted per property per year. Beyond the first 5,000 square feet, the rebate is $1 per square foot.

To date the program has rebated more than $134 million to residential and commercial customers. That accounts for more than 125 million square feet of turf removed. The program's primary funding is new connection charges.

For more information, check out http://www.snwa.com/html/cons_wsl.html.

Richard Atwater also likes the look—and the savings of money and water—that come with losing the traditional lawn. He's head of the utility serving Southern California's Chino Basin, where outdoor water use in the arid climate can account for 50 to 70 percent of total water use. At his home near Pasadena, California, Atwater says he replaced his traditional lawn with low-water landscaping and cut his water use by 50 percent!

"We all have to become smarter about how we use water," says Atwater. "If you go back 200 years in Southern California . . . water was critical to developing every community. When they had only a short supply, those people were really smart about how they used water. Then 50 years ago when the aqueduct that brought water to Los Angeles was built, they said, 'Here it is!' and they gave it away almost free. Guess what? We used it because it was cheap and plentiful, and we made Los Angeles into an oasis. But this is a desert climate.

"Maybe we ought to go back to our roots of 100 years ago when we didn't have cheap water from Northern California or the Colorado River," adds Atwater. "Remember, you can live in a gorgeous place like this and use half the water."

Conservation is an absolutely crucial element of maintaining the water supply at a sustainable level, says Florida water attorney Roger Sims. It can't be an occasional or seasonal idea, agrees Jeff Kightlinger, the Southern California water wholesaler. "This is a lifestyle issue," he says. "We don't have enough water."

"You wouldn't think we'd have these problems in a water-rich state like ours," says Lee Breckenridge, the Northeastern University law professor in Boston. "But many municipalities around Massachusetts must impose water restrictions in the summer because they are facing a water shortage. One of the main problems has been lack of basic conservation measures and overuse of water for lawns. It's like a God-given right to have a lawn."

Water conservation is itself a resource, adds Idaho rancher and former legislator Bruce Newcomb. "It's a way to take water, the resource, and use it more wisely. You have a responsibility to use what you have wisely. I think that's definitely an important part of water policy in any state."

Water conservation needn't always be mandated by government, either. When faced with a drying up of water, carpet manufacturers in Dalton, Georgia, joined forces to conserve that precious commodity. Their cooperation wasn't government-mandated, but they did it anyway.

Water Reuse

Reuse and recycling of water is a no-brainer when it comes to easing the stress on municipal water supplies. El Paso's Archuleta attests to it. (He recently was named a WateReuse Person of the Year 2010 by the WateReuse Association.) Atwater applauds it, too. "We have 18.5 million people from the Mexican border to Ventura County, and we're not going to get any more water from the Colorado River or Northern California," he says. "We have to live within our means."

CONSERVATION INCENTIVES

Many water utilities and municipalities across the country sponsor various water-saving initiatives and conservation education programs for residential and business customers. Check with your local water provider to find out what's available in your area. Some options may include:

- Rebates for high-efficiency clothes washers, dishwashers, and toilets
- Rebates or vouchers to subsidize the use of low-water landscaping or the removal of high-water-use turf
- Rebates for irrigation systems that conserve water
- Incentives for use of synthetic turf—50 cents per square foot, for example
- Rebates to encourage lower water usage
- In-school conservation education programs
- Free landscape and water-use audits

Somehow, says Atwater, his utility must "create more water in Southern California. That means we're going to have to recycle and reuse instead of dumping highly treated wastewater into the ocean. We're going to irrigate our city parks, schools, golf courses, and we're going to use our highly treated wastewater to do it. There's a lot of opportunity to squeeze and become much more efficient in the way we use water."

The state of Florida already has a law requiring part of its sanitary waste to be reclaimed and used elsewhere rather than simply discharged as gray water into the ocean, adds Orlando water attorney

Roger Sims. He cites the Tampa Bay area in west-central Florida and the St. Johns River area in northern and east-central Florida as regions that are doing a good job of reusing water. South Florida, on the other hand, has a way to go, he adds.

WATER TALES

Have a drink of (waste) water! That doesn't sound too appetizing, but the technology exists to make it a reality. Politics and perception may intervene, however, says Orlando attorney Roger Sims. The perception is that it's okay to return treated water to the natural environment— into a river or stream, for example—or to inject it into an aquifer and then recover it somewhere farther down the line. Particularly with ultraviolet and other finishing techniques, the water coming out of state-of-the-art plants is as good as you can get anywhere. But people don't want to drink it.

El Paso's Ed Archuleta agrees. The technology exists to treat wastewater and return it to a drinking water standard for direct potable reuse. The astronauts are doing it, but the country isn't quite ready. There are still many unknowns, and much of the resistance is psychological.

A recognized leader in water reuse and recycling is the Orange County Water District (OCWD) that serves the water needs of nearly 2.4 million people in Orange County, California. Despite what Archuleta suggests—that Americans may not be ready for "toilet to tap" when it comes to their drinking water—OCWD overcame much of the negative public perception with its Groundwater Replenishment System, the largest indirect potable (water) reuse project in the world. Every day 70 million gallons of highly treated wastewater is purified and then reinjected back into the ground—a portion of it is used as a barrier to seawater intrusion, and another portion of it replenishes deepwater aquifers. Water from the Groundwater Replenishment System exceeds all state and federal drinking water standards, too. (You can learn more about how OCWD's efforts do make a difference at http://www.ocwd.com or http://gwrsystem.com.)

To learn more about water recycling and reuse and some of the many success stories across the country, check out the following:

- **U.S. Environmental Protection Agency, Region 9 Water Program:** "Water Recycling and Reuse: The Environmental Benefits," http://epa.gov/region9/water/recycling/
- **WateReuse Association:** http://watereuse.org, or the nonprofit industry organization's new community outreach web site, http://athirstyplanet.com.

Another success story is in South Florida where surface water is being diverted for the Everglades, and reverse osmosis (a purification process of forcing the water through a membrane under high pressure to remove the salts and impurities) is making up some of the demand for new water, says Sims. "The city of Hialeah, in partnership with Miami–Dade County, is building a plant that takes brackish groundwater that's available in abundant supply, filters it, and makes it available for potable use. The salt brine [left over from the process] is sent to an injection well and goes back down below, where it came from."

WHAT ABOUT LAND DEVELOPMENT?

As you've read, how we develop land—with impervious surfaces such as asphalt, concrete, and the like—exacerbates the need to replenish aquifers by paving out the Earth's natural ability to replenish its water supplies. The drenching rain turns to a flood that runs off or parking-lot lake that evaporates, instead of returning the water to the ground. But is it plausible to reconsider the ways land is developed? How about an emphasis on leaving room for replenishment, rather than paving vast expanses of land?

Impervious No More

Some cities, towns, and states have begun to opt for more water-friendly surfaces instead of traditional impervious pavement. Subdivisions and commercial development also can be designed with water sustainability in mind. We mentioned some initiatives in

Chapter 3, including creating vegetated swales instead of concrete ditches along roads and highways, urban landscaping, rain gardens, and riparian buffers.

Another creative approach to solving the pavement problem is the use of pervious rather than impervious pavement. Whether concrete or asphalt, pervious paving material allows water to seep through back into the ground instead of draining off and away (see Figure 8.3). Use of pervious materials boosts the cost of a paving project—10 to 20 percent, according to the National Ready Mix Concrete Association. As yet, that's a big consideration with today's recession-weary, budget-strapped cities and states.

The success of such surfaces also depends on the type of soil involved, says Katherine Austin, AIA, a Sonoma County, California-based architect and former mayor of and planning commission member for Sebastopol, California. Pervious surfaces are not effective with nonporous soil like adobe, she adds. The level of the water table in the area also is a consideration, because nothing will soak into the ground if it's already saturated.

Nonetheless, pervious pavement is one more effective option that can make a difference when it comes to our dwindling water supplies.

WATER TALES

The Stormwater Research and Demonstration Park at Villanova University in Villanova, Pennsylvania, uses pervious concrete and infiltration beds as a solution to the area's flooding and erosion problems. See Figures 8.2 and 8.3 to see how it was accomplished.*

The 1.3-acre drainage area between two dormitories used to be paved with standard asphalt and utilized a conventional storm sewer system that sent the water runoff directly to the headwaters of Mill Creek. That contributed to flooding problems that led to stream bank erosion and sedimentation.

*Villanova University Pervious Concrete Site, Executive Summary (June 19, 2007), http://egrfaculty.villanova.edu/public/Civil_Environmental/WREE/VUSP_Web_Folder/PC_web_folder/PC_main.html.

FIGURE 8.2 The Stormwater Research and Demonstration Park at Villanova University (Villanova, PA)

Source: Villanova Urban Stormwater Partnership.

FIGURE 8.3 Water Draining through This "Puck" Made of Pervious Pavement from the Portland Cement Association

Source: Portland Cement Association.

In 2002, it was rebuilt as a storm water research facility with infiltration beds to filter the runoff, and then paved over with pervious concrete. The redesign's goal was to capture and infiltrate water from storms of up to two inches of rainfall so that there would be virtually no runoff.[4]

MORE WATER-WISE INFORMATION

- ConcreteThinker, from the Portland Cement Association, http://concretethinker.com
- U.S. Environmental Protection Agency's Green Infrastructure program, http://cfpub.epa.gov/npdes/home.cfm?program_id=298
- Invisible Structures, http://rainstore.com
- Porous Asphalt from the National Asphalt Pavement Association, http://hotmix.org/index.php?option=com_content&task=view&id=359&Itemid=863
- Pervious Pavement and Concrete Answers from the National Ready Mixed Concrete Association, http://perviouspavement.org; http://concreteanswers.org

---

Built-In Water Efficiencies

Water efficiency is a key component of the National Association of Home Builders (NAHB) National Green Building Program (http://www.NAHBgreen.org), according to Kevin Morrow, program manager for NAHB's National Green Building Standard. "Energy efficiency historically has been what most people key in on with green building," says Morrow. Until recently, that is, he adds. As people are seeing their lakes dry up, they're taking a more holistic approach to green.

The National Green Building Program encompasses energy and material resource efficiency, indoor air quality, and outdoor green practices such as preserving trees and other native vegetation, as well as reducing outdoor impervious surfaces. The water-efficiency portion of the program includes installation of water-efficient fixtures and appliances like low-flow showerheads, faucets, and toilets; rainwater collection systems; wastewater treatment systems; and hot water recirculation systems.

Morrow adds that factoring in basic green approaches doesn't have to cost big bucks. It can add as little as 1.5 to 2 percent to the cost of a house, according to a study by the NAHB Research Center.

The International Green Construction Code is another industry effort that addresses water, including site development and land use, water resource conservation, rainwater collection and distribution systems and the recovery of used (gray) water, in addition to other green issues. It's a sustainability initiative from the International Code Council, American Institute of Architects, and ASTM International (formerly the American Society of Testing and Materials).[5]

Toilet Talk

One of the biggest water guzzlers in a home is the bathroom. But plenty of water savings are possible, and they've gone high-tech. The drop-the-brick-in-the-toilet-tank-to-use-less-water approach is long gone.

Americans waste nearly 640 billion gallons of water per year flushing old, inefficient toilets. That's the equivalent of 15 days' worth of flow over Niagara Falls, according to data from EPA's WaterSense water savings program. By replacing older, inefficient toilets with WaterSense-labeled low-water-use models, a family of four could reduce water used for flushing by more than 60 percent and, depending on local water and sewer costs, save close to $100 annually.

On Super Bowl Sunday, if every viewer watching the big game flushed at halftime, it would require about 300 million gallons in near sync. On the other hand, if those viewers used water-saving toilets with EPA's WaterSense label, flushers would use about one-third less water.*

*U.S. Environmental Protection Agency, "WaterSense: Statistics and Facts," http://epa.gov/watersense/news/facts.htm; U.S. Environmental Protection Agency, *The WaterSense Current*, no. 9 (Winter 2009), http://epa.gov/watersense/news/current/winter2009.htm#answer.

One of the hottest new water-saving devices is the dual-flush toilet—flush one way for less water to wash down less waste, flush the other way for more water to flush away more waste. "It's a guilt-free kind of toilet," says architect Austin. She spent $275 for her model, and figures that depending on her water charges, she'll save plenty of water and cash in the long run.

WATER TALES

Forget worrying about the wasted water that flows down the drain while your shower or bath water heats up. Katherine Austin, former mayor of Sebastopol, California, has the solution:

What I've done in my home is install a water-recycling pump in the bathroom. The pump goes under the sink in the bathroom farthest from your hot water heater. When you're ready to turn on the hot water, push the button and the still-cold water from the hot tap is pumped back into the cold water pipe and back to the hot water heater, as opposed to going down the drain. When the pump senses the hot water has arrived, it turns off automatically. Then just turn on the faucet and you have hot water.

It's the best thing since sliced bread in terms of saving water in homes because almost every older home has that problem.

Austin noted that the pump cost her less than $200. Pump models are available for $180 to $400, work with standard and tankless water heaters, and are easy to install.

CRACKDOWN ON POLLUTION

No matter where you live, chances are more than once you've rejoiced from that rain that finally broke the heat spell or that downpour that cleaned the air. That same storm also washed away the dirt, grime, pollution, and more that collects on streets, walkways, buildings, cars, trees, and everything else that makes up the landscape—urban or rural. That gunk, grit, and pollutants in turn likely end up in someone's source water somewhere! Nonpoint source pollution or NPS (also known as storm-water runoff), whether from farm fields or urban landscapes, taints much of the water we have.

"Storm water runoff is one of the major water quality and quantity issues," says Philadelphia attorney Ken Warren. Runoff often carries toxic and nontoxic pollutants into streams and causes degradation. Increased runoff rates and volumes of water from developments that create impervious surfaces also impair streams. Unfortunately, existing regulation of storm-water runoff frequently is inadequate, he adds. The federal Clean Water Act permitting system controls only discharges from point sources like pipes and conduits. That encompasses some municipalities' separate storm sewer systems, construction, and industrial activities.

Warren would like to see national legislation considered that would give the EPA greater authority to manage storm water. "It could very well be administered by state or regional agencies. But some additional arrows in the quiver of the water managers are definitely needed," he says.

The pocketbook incentive. Some cities and utilities are trying to influence consumers toward a greener approach to development and decreased runoff pollution by introducing tiered storm-water fees. The greater the area of impervious surface on a property, the higher the storm-water runoff fees. The Sierra Club's Ken Kramer likes this financial incentive approach as long as there's a lifeline (discounted) rate for low-income consumers. "Any time you can use the cost factor in developing different approaches, you have a better chance of succeeding in your ultimate goal," says Kramer. "Some financial incentive to do the right thing or disincentive to do the wrong thing is needed to push people to change their behavior."

Of course, the cash incentive works both ways. Consider Florida's deal with U.S. Sugar to buy tens of thousands of acres of land in South Florida as part of a plan to clean up the Everglades.

LIMIT RUNOFF POLLUTION

The EPA suggests several ways in which the average person, no matter where he or she lives, can lessen the impact of water runoff, known as nonpoint source pollution, or NPS:*

- Keep litter, pet wastes, leaves, and debris out of street gutters and storm drains. These outlets drain directly to lakes, streams, rivers, and wetlands.
- Apply lawn and garden chemicals sparingly and according to directions.
- Dispose of used oil, antifreeze, paints, and other household chemicals properly, not in storm sewers or drains. If your community does not already have a program for collecting household hazardous wastes, ask your local government to establish one.
- Clean up spilled brake fluid, oil, grease, and antifreeze. Do not hose them into the street, where they can eventually reach local streams and lakes.
- Control soil erosion on your property by planting ground cover and stabilizing erosion-prone areas.
- Encourage local government officials to develop construction erosion and sediment control ordinances in your community.
- Have your septic system inspected and pumped, at a minimum, every three to five years so it operates properly.
- Purchase household detergents and cleaners low in phosphorus to reduce the amount of nutrients discharged into lakes, streams, and coastal waters.

*U.S. Environmental Protection Agency, "Polluted Runoff (Nonpoint Source Pollution): What You Can Do to Prevent NPS Pollution" http://epa.gov/owow/nps/whatudo.html.

GOVERNMENT IN THE PICTURE

Government must play a role, too, at the local level and up. Remember the contaminated water at Camp Lejeune, North Carolina? The government long had been accused of dragging its feet on that investigation. Guilt or innocence aside, the case is a prime example of what can and does happen. We've talked about other water conflicts across the country that drag on for various reasons—the parties involved, the legalities, the approaches, the problems, or the high stakes in today's water game.

When you run out of resources, you must look at things differently, says Sandia Laboratories' Hightower. The United States is in that position right now. "We're having a hard time figuring out how to get a handle on this water issue, especially when we have more than 20 different federal agencies managing water," he says. "Someone is going to have to bite the bullet and say, 'Okay, we're going to do this differently somehow and manage this more appropriately.' There are many different ways to do that. But how are we, as a big country, going to actually move in that direction? It may take another water crisis or a huge drought. Unfortunately, that's always too late to do any worthwhile planning."

"I'm quite optimistic that we are dealing with the problem at a time when we still have adequate water supplies to meet our needs if we husband the water in the proper ways," adds Clive Strong, Idaho's deputy attorney general.

WATER TALES

 Cooperation and conservation work when it comes to making the most of America's water resources. Recognizing that, the U.S. Department of the Interior (DOI) annually honors success stories with its Cooperative Conservation Award. (For more on cooperative conservation, check out the DOI's partnerships Web site: http://www.doi.gov/partnerships/ccawards .html.) Here are two recipients who have achieved real water results.

- The Deschutes River Conservancy (DRC) in Oregon is a nonprofit corporation bringing together federal, state, American Indian, and local governments with private stakeholders to carry out basin-wide ecosystem restoration projects. Among its accomplishments:
    - During the 2006 irrigation season, DRC projects increased summer flows in the Middle Deschutes to record levels,

achieving the 100-cubic-feet-per-second milestone for the first time since irrigators began diverting water from the river in 1899.

♦ The DRC implements a water conservation program and a water transfer program in the form of water banks that are a valuable resource for irrigation districts in Central Oregon. For example, through the Central Oregon Water Bank, the DRC facilitates transfer of water from irrigation districts to meet the needs of Oregon communities and the Deschutes River.*

• South Arkansas' Sparta Aquifer Recovery Initiative and its public and private partners in southern Arkansas and northern Louisiana have achieved massive reductions in water withdrawal from their stressed aquifer. Water levels had dropped more than 360 feet near El Dorado, Arkansas, before the conservation project got underway. Today aquifer withdrawals are down about 7.5 million gallons per day and well levels have risen.[†]

*U.S. Bureau of Reclamation, press release, "Deschutes River Conservancy to Receive Interior's Cooperative Conservation Award," May 9, 2007, http://www.usbr.gov/newsroom/newsrelease/detail.cfm?RecordID=16861.

[†]U.S. Department of the Interior, "Citation: Cooperative Conservation Award, South Arkansas Sparta Aquifer Recovery," http://cooperativeconservation.gov/awards2008/ccawardcitationSoArkansasSparta.pdf.

California as Leader

The state of California exemplifies the role government can play to promote water-sustainable practices in construction. Through its Water Quality Control Board, it mandates a standard urban storm water mitigation plan (SUSMP) that specifies how developers must deal with recharging aquifers and limiting runoff pollution.

Having an SUSMP has forced construction practices in the state to change, says architect Austin. "When we do a development now, we have to prevent any additional runoff from what existed previously. The development has to be able to store and cleanse storm water before it goes into the natural drainage system or the storm drain. Imagine [that] you have an empty field in which all the water gets absorbed. If you develop it with roads, houses, driveways, and sidewalks, you can't add anything to the runoff."

That means all the rainwater from the downspouts, off the roofs, and running on the pavement has to be stored and cleaned—in the case of some pollutants, removed—before it goes back into the gutters as runoff. That's a tall but doable order, helped by several techniques including using sand, trees, and grassy areas as filters, says Austin. Downspouts aren't connected directly into storm drain systems, either. Instead, they come down to the bottom of a house and in general pour onto gravel, rock, and natural grasses—not turf—that naturally slow down the water flow and cleanse the water. Another drain system then may dump the water into a large planter strip between the curb of the street and the sidewalk. The native grasses in the planter absorb much of the water, cleanse it, and then allow it to flow through to the lowest point in a subdivision, where the cleansed and slowed-down water then flows into the drainage system.

Such environmental awareness comes at a price, however, because it requires more space, which means less room to build homes. "The Water Quality Control Board has legal status in our state and can fine or arrest people if they're violating water quality laws," says Austin. "You can't get a subdivision here approved unless you are meeting the requirements of [the] laws, so everybody is on an equal playing field. Every builder has to do it. They know everybody else is spending the money or devoting the land to do it, so they all do it."

Implementing that kind of legal requirement is the way to get water-conscious construction techniques implemented elsewhere in the country, adds Austin. "It's not going to be popular. But it can be done, and it has to be done. This can be done city by city, but it's far more effective on a state or regional basis."

YET ANOTHER BLUEPRINT

The Subcommittee on Water Availability and Quality of the National Science and Technology Council's Committee on Environment and Natural Resources in 2007 proposed a strategy to deal with America's water crisis.* It laid out seven factors that must be addressed:

1. Implement a national water census

2. Improve hydrologic prediction models and applications

3. Develop collaborative tools and processes for U.S. water solutions

4. Develop a new generation of water-monitoring techniques

5. Develop and expand technologies for enhancing reliable water supply

6. Develop innovative water-use technologies and the tools to enhance public acceptance of them

7. Improve understanding of the water-related ecosystem services and ecosystem needs for water

*Committee on Science and Technology, Subcommittee on Energy and Environment, U.S. House of Representatives, "A National Water Initiative: Coordinating and Improving Federal Research on Water," June 23, 2008, http://gop.science.house.gov/Media/hearings/energy08/july23/charter.pdf; National Water Research and Development Initiative Act of 2009, summary of major provisions, http://thomas.loc.gov/cgi-bin/cpquery/?&sid=cp111alJsu&refer=&r_n=hr076.111&db_id=111&item=&sel=TOC_36907&.

INDIVIDUALS CAN MAKE A DIFFERENCE

When it comes to water use, individuals can make a difference. If in a single day everyone across the country saved one gallon of water the total savings would be a quarter-billion gallons of water.

Still not convinced that each person can make a difference? Remember the U.S. Geological Survey (USGS) Drip Accumulator we talked about in Chapter 3? One dripping faucet does add up to a big deal (http://ga.water.usgs.gov/edu/sc4.html). One faucet that drips 60 times per minute at an estimated ¼ milliliter per drip (USGS's estimate) loses 86,400 drips, or 5 gallons, per day and 2,082 gallons per year. Multiply that by the number of households in the United States—105,480,101 in 2000, according to U.S. Census numbers—and that's an amazing 219.6 billion gallons of water wasted every year. Of course, not every household in the country has a drippy faucet. Or does it?

INFORMATION SOURCES

Plenty of free information on water and water issues is only a click away. A good starting point is the USGS's Water Science Center. To find information for your state, visit http://www.usgs.gov/state/.

Alternatively, check out the EPA's Local Drinking Water Information site, http://epa.gov/safewater/dwinfo/index.html, and click on your state.

Most states also have water-specific web sites.

The EPA even has a WaterSense program that promotes water savings via specially labeled water-efficient products (http://epa.gov/watersense/). WaterSense-labeled faucets, for example, are 30 percent more efficient than the standard bathroom faucet. Translation: the average family could save nearly 600 gallons of water a year, enough to do 10 loads of laundry. Water savings save you cash, too. If every household in the United States installed water-efficient appliances, the country would save more than 3 trillion gallons of water and more than $18 billion dollars per year, EPA WaterSense estimates.

Here are some more WaterSense suggestions:

- Turning off the tap while brushing your teeth can save eight gallons of water a day, or almost 3,000 gallons a year.
- Allowing the faucet to run while washing or shaving is a big waster of energy, and that energy takes water to produce. Running the faucet for five minutes uses about as much energy as operating a 60-watt light bulb for 14 hours!

Urban and regional planner Daniel Williams suggests a new, and what he calls "conservative," approach to our nation and its water:

"'Radical' is putting up a huge concrete embankment to stop a river that's half the size of our continent. 'Conservative' means to look at things as they are and figure out how to work with them. Radical is what the approach has been. That's not to say that was always the wrong choice, but it is now. True sustainability comes from understanding the very large picture and acting at that scale."

THE NEXT STEP

A lot of wishful thinking runs through the water business, says water negotiator Steve Mumme. "That's why I say planning by disaster is a significant part of the planning process. It's a very human drama. The things that are happening now are reshaping lives. Some of it involves learning that must occur, and some of it is just damned unfair."

Wherever anyone stands on the water issues, on climate change, on shortages, abundance, planning, rights, conservation or lack thereof, politics, and more, is not as important as the realization that change is inevitable.

Utah farmer Jerald Anderson takes the pragmatic approach, despite the threat that he may lose the water essential to his current livelihood. He even suggests that history might not judge today a "crisis" in terms of water supplies.

"Water has always been a problem in the West, particularly as we've improved agriculture efficiency. We've been able to do things we weren't able to do before, and we've used water to do it. We've used energy to accomplish things we just couldn't have accomplished any other way, too. But in terms of the word 'crisis,' I think you have to carefully consider the time scale," he says. Because of high prices on something or dwindling supplies, we as a nation might believe a resource is at a "crisis" point, whereas in the long term, says Anderson, "it may never even make the list of problems."

"Yes, it may be an individual crisis," he adds, "but as a society or a country it may not be. It may be that it forces us to adjust. Frankly, no society should expect that things are always going to be the way they are today. If we plan in that direction, we're doomed to failure. You have to know that things will change in the future, and you must be willing to accept that change."

As you've seen throughout this book, more people are recognizing that learning—and change—must occur. The head-in-the-sand approach to water won't work anymore. We simply can't take this clear-gold twenty-first-century commodity for granted any longer.

"We can't just think of water as something we need in terms of something flowing out of our tap when we turn on our faucets," adds the Sierra Club's Ken Kramer. "We need to think of it more comprehensively, as something that's needed for both people and the environment, and that environmental water needs also are important to people."

You can make a difference. Every drop does count.

WATER REALITIES

- Individuals, businesses, and communities all can and do make a difference with conservation, concern, and awareness of water issues.
- Experts urge Americans to reevaluate how they use water and to think in terms of needs versus wants.
- A major approach to saving our precious water resources involves planting native grasses and xeriscape instead of water-consuming turf. Some utilities even offer cash rebates to customers who replace their lawns with low-water-use landscaping.
- Other ways to combat big water wastes across the country include fixing leaky faucets and replacing older toilets with new, low-water-use models. One leaky faucet or one old water-guzzling toilet does make a difference when multiplied by the millions of households across the country.
- Nonpoint source pollution (NPS) or storm water runoff is a major polluter of our water supplies. Efforts to cut down on NPS include using pervious (porous) surfaces as opposed to solid concrete and asphalt, changing approaches to drainage, and raising community awareness of the problem.
- No matter what their politics, personal positions, or geographic location, Americans must face the facts: Water no longer is an infinite resource. We all must recognize that how we have used the resource in the past isn't necessarily the way of the future.

NOTES

1. David E. Stooksbury, "Mild drought returns to parts of Georgia," http://georgiafaces.caes.uga.edu/index.cfm?public=viewStory&pk_id=3886; David E. Stooksbury, "Drought Covers Most of Georgia," http://georgiafaces.caes.uga.edu/index.cfm?public=viewStory&pk_id=3979.

2. U.N. Food and Agriculture Organization, FAO Newsroom, "Coping with Water Scarcity: Q&A with FAO Director-General Dr. Jacques Diouf," March 22, 2007, http://www.fao.org/newsroom/en/focus/2007/1000521/index.html; "FAO Urges Action to Cope with Increasing Water Scarcity," March 22, 2007, http://www.fao.org/newsroom/en/news/2007/1000520/index.html.

3. John Bredehoeft, Jon Ford, Bob Harden, Robert Mace, and James Rumbaugh, "Review and Interpretation of the Hueco Bolson Groundwater Model," report prepared for El Paso Water Utilities, March 2004, http://www.epwu.org/water/hueco_bolson/ReviewTeamReport.pdf.

4. Villanova University Pervious Concrete Site, Executive Summary, June 19, 2007, http://egrfaculty.villanova.edu/public/Civil_Environmental/WREE/VUSP_Web_Folder/PC_web_folder/PC_main.html.

5. International Code Council, press release, "New Green Construction Code Unveiled," March 15, 2010, http://www.iccsafe.org/cs/IGCC/Documents/NR/NR031510-IGCC-Unveiled.pdf.

EPILOGUE

Get used to it. Aqua shock, whether it's hit you personally yet or not, is the way of life in this twenty-first century. *Aqua Shock* makes that clear. In these pages, I've presented some of the issues, problems, and concerns swirling around the nation's water. The particulars of what's threatening or ailing your personal water supply depends on where you live, what you do, the laws of your locale, and how you live your life. The bottom line, though, is the water equation:

Growing demand + Limited supply = Shortage and conflict over what's left

What happens if we run out of water? The lessons of history tell us that when a civilization exhausts its water supply it disintegrates. Colorado-based paleohydrologist Kenneth R. Wright—founder and chief engineer of Wright Water Engineers (http://wrightwater.com)—points to settlements like Chaco Canyon in northwestern New Mexico and the Anasazi in southwestern Colorado. Both societies, he says, likely collapsed in part as a result of civil unrest caused by lack of water.

Today we live in a global society. If the water runs out, we can't just migrate elsewhere. And there's no silver bullet—no single solution—to the nation's or the world's water woes. The plain fact is that we don't yet fully realize all the problems, let alone the solutions.

No matter where you live, whether it's wet or dry, and no matter your personal agenda, we as citizens and civilizations cannot afford to exhaust this finite and clearly precious resource. In 2010, we've already seen modern-day unrest as a result of water shortages—on the other side of the world in places like Pakistan (well before the devastating floods of summer 2010); in India, China, the Philippines, Sri Lanka, Nepal, and beyond (remember the 30th parallel conundrum discussed in Chapter 2). Google "water shortage" and "riots" and thousands of results come up. Closer to home, California farmers joined by then-Governor Arnold Schwarzenegger protest shriveled fields. No one—at least publicly—seems to have drawn guns, but the fights are vicious nonetheless—in the courts, legislatures, councils, Congress, and on the protest lines.

Solutions begin with recognition of the problems. You've read how some scientists, individuals, groups, governments, and others are paying attention; they recognize the issues and are making a difference. *Aqua Shock* is the wake-up call for all of us to pay attention to our planet's diminishing resource of fresh water. The alternative is unthinkable.

THE GREAT JOB GENERATOR

The current water outlook may be muddled, but it figures clearly and prominently in our recession-weary economy. *Water*, and all things related to it, will be the great job generator of the twenty-first century. With so many aspects of water sources and systems outdated, in disrepair, or just wrong—from approach, to infrastructure, monitoring, delivery, analysis, intervention, use, alternatives, laws, and beyond—it's impossible for water-related needs not to generate hundreds of thousands, likely millions of jobs and opportunities for the future.

Fixing our nation's crumbling water infrastructure alone is a mega-billion-dollar—if not trillion-dollar—proposition. Hundreds of thousands of miles of water pipes crisscrossing below the surface in the United States are worn out and need replacing and/or repair. Day after day the reports of water main breaks signal a tiny fraction of the problem. Creativity will rise to the surface, too. Someone had a better idea: Develop a product to reline the pipes rather than replace them. Voilà! A new idea, a new industry, new jobs. No doubt many other new "better" ideas will open the flood gates to change.

Opportunities abound, too, for water conservation and cleanup—perhaps new and better techniques, invention, or the manufacture and sale of materials related to preserving and restoring available resources. The city of Malibu, California, recently dedicated its Malibu Legacy Park, 15 acres in a central park that's also a state-of-the-art storm water and urban runoff facility designed to clean up the water (http://www.malibulegacy.org/media.html). It's a facet of that city's publicly expressed $50 million commitment to clean water.

More jobs will come from other solutions—high-tech or low-tech—in areas like water utilization, purification systems, desalination, clean-up, drilling, analysis, sustainable construction, and more. The list of water-related needs in the twenty-first century is literally endless—as are the opportunities for entrepreneurship and creative solutions. Gleaning water from thin air, as in atmospheric water generators—machines that capture moisture from the air, purify it, and deliver it on call—isn't far-fetched. It's a reality from companies like EcoloBlue (http://www.ecoloblue.com) and WaterMicron (http://watermicron.com).

An Israeli company, Arad Group (http://arad.co.il/news.asp), has even developed a small military-like, GPS-guided drone that reads water meters and detects water leaks from the sky. A microbiologist in South Africa (he's also dean of Science at Stellenbosch University) with others has created a small water filter dubbed the "nano tea bag" that fits into the neck of a bottle to purify water (http://blogs.sun .ac.za/news/2010/07/28/su-scientists-develop-a-high-tech-%e2%80% 98tea-bag%e2%80%99-filter-for-cleaner-water/).

On a more low-tech level, Wisconsin-based Second Rain Inc. (http://www.second-rain.com) has a do-it-yourself rain catchment system that has a rainwater diverter that attaches to a downspout and a wooden bench that actually holds 65 gallons of rainwater. (Best check if catching water that falls from the sky is legal in your state, first!)

Remember Greg Huff, from the Iowa Rural Water Association mentioned earlier in the book? He and his wife and kids decided to do something about the fact that, elsewhere in the world, people are dying for lack of clean water. The family launched the charity ICanSaveLives (http://icansavelives.org), which raises money by collecting empty pop cans and returning them for recycling. The money raised is used to provide water purification and filtration systems, hand washing stations, sanitary sewers or latrines, and hygiene education to villages, hospitals, schools, and other small public facilities in developing nations.

So what happens when our water runs out? It doesn't have to. Each of us can make a difference. Today's legacy to future generations doesn't have to be a dried-up Earth. We can start now to save our planet's water. It's up to each and every one of us.

WHERE AND HOW TO START

The problems may be complex, but not all solutions must be. Here are a few simple things that anyone can do to get started making a difference in our nation's water supply. Better still, some of them even save you cash. Some water-wise thoughts from various people and organizations, including the EPA:

- Get rid of old, inefficient appliances and water fixtures; EPA says it will save you 30 percent water use every day.
- Americans waste nearly 640 billion gallons of water every year flushing old, inefficient toilets. That's 15 days' worth of water flow over Niagara Falls.
- Fix the drips—remember the "drip accumulator."
- Turn off the water when you're not using it. Just doing so while brushing your teeth can save up to eight gallons of water a day per person. Multiply that by 1 million people and the numbers very definitely add up quickly—8 million gallons saved every day, with 365 days in a year—that's almost 3 billion gallons of water saved a year just by turning off the water while brushing your teeth!
- Take one less bath a week. That's about a 50 gallons-per-person savings depending, of course, on the size of your bathtub. If the same 1 million people bought into the idea, it could save another 2.6 billion gallons of water a year.
- Take your car to a professional car wash instead of hosing off that dirt and grime (pollutants) in your driveway. Many car washes recycle or at the very least clean pollutants out of the water before it ends up back in our water supply.
- Don't toss those old, outdated pills down the drain. Drop them off at a designated "unwanted drugs" collection facility

instead (ask your pharmacist for locations). Our nation's wildlife will thank you.

- If you're using a private well for your water supply, have it tested for contaminants. It's a small price to pay—some places may offer free kits—to prevent serious health problems from invisible contaminants.
- Properly dispose of hazardous wastes, including used motor oil or unused gasoline from your lawn mower or snow blower tanks. That helps keep contaminants out of our source water.
- If your locale doesn't have outdoor lawn-watering restrictions in place, consider limiting your water use anyway. It saves water, and puts less strain on the system. (Better still, consider replacing your lawn with low-water native plants, especially if you live in an arid or semi-arid climate.)
- Pay attention to posted and/or other notices and warnings related to water and water supplies. The bottom line is safety.
- Don't waste water. That's our shared responsibility.

LAST WORDS: If you have a way to make a positive difference when it comes to our planet's water supply, tell me: http://www.susanjmarks.com.

GLOSSARY

acre-foot. Typical measure of the amount of water when referring to large quantities; the equivalent of one foot of water covering one acre of land.

adjudication. Legal proceeding to examine the issues pertaining to, and then determine the outcome of, a dispute.

aqueduct. A channel, canal, pipeline, or other similar means to transport water.

aquifer. Water that is found or stored, sometimes moving, underground in generally porous or loose rock, sand, or gravel formations that are saturated with water. The top layer of the aquifer is known as the water table.

arid *or* **semi-arid.** Dry climate lacking in moisture.

artesian water. Water that is under pressure so that it rises to the surface on its own.

atmosphere. The layer of gases surrounding Earth.

beneficial use. In prior appropriation doctrine, a water right is granted for beneficial use only; open to interpretation.

brackish water. Saline water; can occur in aquifers as a result of overpumping or draining down the aquifer; may also occur naturally, depending on the origin of the water.

call for water. To demand delivery of an allocation, in this case in accordance with a right to a certain amount of water.

climate change. Any major change in climate—including precipitation or temperature—lasting for an extended period.

condensation. What happens when water changes from gas to liquid.

curtailment. In water rights, stopping a junior rights holder from taking an allocation of water in order to fulfill the allocation demanded by a senior rights holder.

dead pool. The level of water in a reservoir below which hydroelectric power cannot be generated and no more water can be released downstream.

depletion. Using up of supply; with water, its removal without replacement.

deposition. When matter changes from a gas to a solid without passing through the liquid phase; the opposite of sublimation; in the case of water, when water vapor turns to ice or snow.

desalination. The process of removing salt from brackish water or seawater.

drainage basin. Watershed or portion of land that drains the rainfall and water of the area to one outlet, such as a reservoir.

drawdown. Lowered level of underground water supply; usually caused by overpumping.

drought. Moisture deficit over a broad area, and usually over a prolonged time, that results in shortfalls in the water needed to meet the needs of people, business, and industry.

dry-land farming. Relying on rainfall and naturally occurring surface water (as opposed to irrigation) to raise crops.

El Niño/La Niña. Refers to oscillation of the ocean and atmosphere system in the tropical Pacific Ocean that affects weather patterns around the globe; specifically, a change in ocean water surface temperatures. With El Niño, the temperature rises; with La Niña, the temperature cools.

evaporation. The process by which moisture in liquid form changes to a gas.

evapotranspiration. A combination of evaporation and transpiration; refers to the process whereby water evaporates into the air and transpires from plants into the air.

federal reserved water rights. Those rights to water associated with the federal government's reservation of land for public use, such as Indian reservations, military reservations, national parks, and more. In the case of Western water rights law, the date of priority is the date the reservation was established; doctrine established by U.S. Supreme Court in *Winters v. United States* (1908).

futile call. A demand for delivery of a certain quantity of water that is in accordance with a right to it, but that cannot be delivered without unreasonable waste, such as when attempts to fulfill the demand or deliver the water would be futile.

global warming. Average increase in temperature near the surface of the Earth; can be natural or man-made; the latter includes increased emission of greenhouse gases.

gray water. Noncontaminated water that can be recycled.

Great Lakes. The largest freshwater lake system in the world; holds one-fifth of Earth's freshwater.

groundwater. Water found beneath the surface of the Earth; can be just below the surface, shallow, or much deeper. It can be the source water for various supplies.

High Plains aquifer. A prehistoric aquifer that covers 174,000 square miles in parts of Colorado, Kansas, Nebraska, New Mexico, Oklahoma, South Dakota, Texas, and Wyoming; the largest aquifer in the United States.

Hoover Dam. U.S. Bureau of Reclamation dam project on the Colorado River between Arizona and Nevada that created Lake Mead (a major recreation destination); built to control flooding, produce hydroelectric power, and store water; originally called Boulder Dam.

hydrologic cycle. The unending, closed process by which Earth replenishes its water supplies; also known as the water cycle.

ice cap *or* **polar ice cap.** Glacial ice at Earth's North and South Poles that remains frozen year-round.

infiltration. The process of water seeping back into the underground supply; length of time required depends on the amount of precipitation that falls, composition of the soil, frequency of precipitation (degree of moisture in the air), and groundcover, among other factors.

impervious surface. A surface that does not allow water to permeate it.

inorganic contaminants. Impurities that include metals and salts and can occur naturally, or from storm water runoff in urban areas, wastewater discharge, oil and gas production, mining, or farming.

junior water rights and senior water rights. Under water law in most of the western United States, allocations of water are generally based on a system of ownership of the right to use an allotment of water based in part on the concept of first in time, first in right; the first person to put water to *beneficial use* holds the senior (older) right to the water.

Lake Baikal. Lake in Siberia near the Mongolian border; contains about 20 percent of the world's total surface fresh water.

Lake Mead. Created by Hoover Dam on the Colorado River; primary water supply for Las Vegas, Nevada.

microbial contaminants. Impurities that include bacteria and viruses, and can come from animal wastes, sewage treatment plants, and septic systems.

mineral rights. The right to ownership of minerals (oil, gas, gold, etc.) on a piece of land.

mining water. Taking out more water than is replenished; analogous to mining coal.

nonpoint source pollution (NPS). Water runoff from a nonspecific source that is contaminated by various man-made toxic and nontoxic substances, which may range from chemicals and nutrients to sediments and waste.

Ogallala aquifer. Underground water supply that is part of the much larger High Plains aquifer; often used incorrectly to refer to the High Plains aquifer.

organic chemical contaminants. Impurities that include synthetic and volatile organic chemicals; they often are by-products of industrial processes and petroleum production but can also come from gas stations, urban storm water runoff, and septic systems.

Palmer Drought Index. A common numerical measure used to define and predict drought and its various levels of severity.

pervious surface. In terms of pavement, an aggregate that creates a porous surface, allowing water to permeate it.

pesticides and herbicides. Substances used to kill animals/insects and plants. These come from agriculture, irrigation, and/or storm water runoff as well as residential runoff.

polar ice cap. See **ice cap**.

potable water. Water that is considered safe for human consumption.

precipitation. Water that falls from the sky in various forms, including rain, snow, and hail.

prior appropriation. A private property approach to water use that establishes ownership of water rights based on first in time to put water to full beneficial use, first in right to use it.

public trust. The basis for riparian water law in the East; stems from English common law; states that water is for the use of all.

radioactive material contamination. Impurities in the water that can occur naturally or as a result of oil and gas production or mining activities.

recharge. In the case of groundwater, the infiltration or absorption of surface water to replenish groundwater supplies; may be natural or artificially injected.

regulated riparianism/permitted riparianism. System of public property in which the state manages water in trust for the public through time-limited permits.

renewable versus prehistoric aquifer. A shallow aquifer or underground water storage area, versus a deep-water, ancient one that may not fully replenish in a typical human lifetime.

replenishment or recharge. The manner in which water in aquifers is replaced.

reservoir. A man-made lake for storage of water.

reverse osmosis. A purification process in which saline or brackish water is forced under high pressure through a membrane to remove its salt content.

riparian right. The right of a landowner whose property abuts a body of water to reasonable use of that water; originates from English common law.

river basin. A watershed; the geographic area that drains to a common low point.

snowmelt. For purposes of water supply, the amount of snow that melts off in the spring to replenish surface and groundwater supplies; a major source of the U.S. water supply.

snowpack. Deep snow in the nation's high country; eventually melts to produce water supplies.

soil moisture. A measurement of the amount of water contained in soil.

stream flow. A measure of the volume of water in a stream; based on the amount that passes a specific point in a given time period, often expressed in cubic feet per second (CFS).

source water. Water from rivers, lakes, streams, or aquifers that is not treated and is used to provide drinking water via either wells or public water supply.

sublimation. When a substance changes from a solid to a gas without passing through the liquid phase; in the case of water, refers to ice or snow vaporizing without first melting into water.

surface rights. The right to use of the top area of the ground; mineral rights are separate.

surface water. Water found aboveground, not buried, including rain, snow, and ice; usually refers to water in lakes, rivers, streams, ponds, and other bodies of water.

total maximum daily load (TMDL). A measure of the highest amount of a pollutant that a body of water may contain and still meet water quality standards.

transpiration. The equivalent of a plant "breathing"; in this case, releasing water vapor into the air.

underground water. An aquifer; water found beneath the ground; top level of the aquifer is called the **water table**.

wastewater. Water that has been used and discharged as sewage.

water bank. A market-based exchange for the buying, selling, accumulation, and trading of water and water rights.

water district. Entity involved in the administration or disbursement of water.

water right. In water law in the western United States, entitlement to a certain quantity of water, usually in the order of first in time, first in right.

water table. The top level of an underground water source or aquifer.

watershed. Drainage basin or portion of land that drains the rainfall and water of the area to one outlet, such as a particular reservoir, river, or stream.

xeriscaping. Landscaping using native drought-resistant or low-water-use plants and vegetation.

RECOMMENDED RESOURCES

Alliance for Water Efficiency (http://allianceforwaterefficiency.org): Chicago-based nonprofit stakeholder organization that promotes efficient and sustainable use of water, and provides information and assistance on water conservation efforts.

Alliance for Water Quality (http://a4we.org): North American nonprofit advocate for water-efficient products and programs; offers information on water conservation efforts.

American Water Works Association (http://www.awwa.org): This 60,000-member industry organization offers all kinds of education and information on water, from conservation and current events to rates and research. Check out the Water Library.

A Thirsty Planet (http://www.athirstyplanet.com): A community outreach and education web site on the need, uses, and benefits of water reuse and desalination from the nonprofit WateReuse Association.

Centers for Disease Control and Prevention (http://cdc.gov): Find out more about waterborne diseases and their symptoms.

City of Atlanta Department of Watershed Management (http://atlanta watershed.org/drought/): Drought information center for metropolitan Atlanta.

Colorado Compact (http://www.usbr.gov/lc/region/g1000/pdfiles/crcompct .pdf): Read the details of the division of the Colorado River water.

Department of the Interior, U.S. Bureau of Reclamation (http://www .usbr.gov/pmts/water/): Source of information on water reclamation (including desalination) and research.

Environmental Law Foundation (http://envirolaw.org): Find out more about this nonprofit's study of water contaminants.

Environmental Protection Agency (http://epa.gov).

Environmental Working Group (http://ewg.org): Nonprofit that provides research and reports on issues that affect the public health and environment.

Georgia Drought.org (http://georgiadrought.org): The University of Georgia College of Agricultural and Environmental Sciences provides updates and information on drought conditions in Georgia.

Great Lakes Information Network (http://great-lakes.net/lakes): Learn all about the transboundary Great Lakes.

International Bottled Water Association (http://www.bottledwater.org): Industry organization that offers a lot of what you want to know about bottled water, the numbers, and the issues.

Interstate Commission on the Potomac River (http://www.potomacriver .org/cms): Find out the water supply outlook for the nation's capital and learn about conservation.

Local Drinking Water Information (http://epa.gov/safewater/dwinfo/ index.html): Find out about water quality.

NASA Earth Observing System (http://eospso.gsfc.nasa.gov): Learn what's happening on Earth from an out-of-this-world perspective.

NASA Total Ozone Mapping Spectrometer (TOMS) (http://jwocky.gsfc .nasa.gov/aerosols/today_aero.html): Find out what's happening in the atmosphere.

National Association of Homebuilders (http://www.nahb.org): Check out this homebuilding trade association's Green Building Program.

National Ground Water Association (http://www.ngwa.org): Learn more about groundwater here.

National Institutes of Health (http://www.nih.gov): Government agency with information on various health issues, including those arising from contaminated water.

National Integrated Drought Information System (http://www.drought .gov): U.S. Drought Portal provides a snapshot of water conditions across the country.

National Lead Information Center (http://epa.gov/lead/pubs/nlic.htm): A good starting point to learn about lead poisoning and contamination.

National Oceanic and Atmospheric Administration (www.noaa.gov): Government agency of scientists, researchers, and experts on all things weather- and climate-related, including:

    Climatic Data Information Center (http://www.ncdc.noaa.gov/oa/ ncdc.html): Monitors national and world climate developments (including drought) as well as originates the State of the Climate reports.

    Great Lakes Environmental Research Laboratory (http://www.glerl .noaa.gov): Monitors and reports on Great Lakes issues, including water levels, toxins, and forecasts.

    National Weather Service Climate Prediction Center (http://www.cpc .noaa.gov/): Tracks and predicts climate-related developments, including temperatures, drought, precipitation, weather events, and more.

National Weather Service Hydrologic Information Center (http://www
.weather.gov/ahps/hic/index.php/index): Find out what's happening
with the rivers and snowpack in any area.

Natural Resources Conservation Service (http://nrcs.usda.gov/): Part
of the U.S. Department of Agriculture; formerly called the Soil
Conservation Service. Provides leadership to help private landowners
and managers conserve soil, water, and other natural resources.

Natural Resources Defense Council (http://www.nrdc.org): Find out
more about this nonprofit's studies of the nation's water.

New England Interstate Water Pollution Control Commission (http://
www.neiwpcc.org): Not-for-profit interstate agency (member states
include Connecticut, Maine, Massachusetts, New Hampshire,
New York, Rhode Island, and Vermont) that fosters collaboration on
water issues, trains environmental professionals, coordinates water
research, educates the public, and provides leadership in water management and protection.

Office of Water (http://epa.gov/water): Click on Regional Information
to find out about any region's water, or one of many other water
topics.

Pacific Institute (http://pacinst.org): Not-for-profit Oakland, California-
based research institute that emphasizes sustainability and environmental solutions.

ProtectDrinkingWater.org (http://www.protectdrinkingwater.org): Web portal
of the Source Water Collaborative, a 23-member organization group
interested in protecting America's sources of drinking water.

Ridenbaugh Press (http://www.ridenbaugh.com/waterrights/): All about
water rights and adjudications, from Pacific Northwest public affairs
publisher.

Safe Drinking Water Information System (http://epa.gov/enviro/html/
sdwis/sdwis_query.html): Search its state-by-state database for information about public water systems and their violations of EPA's drinking
water regulations.

Sandia National Laboratories (http://sandia.gov): A multi-program laboratory operated by Sandia Corporation, a Lockheed Martin company, for the U.S. Department of Energy's National Nuclear Security
Administration.

Source Water Protection (http://water.epa.gov/infrastructure/drinking-
water/sourcewater/protection/index.cfm): Learn all about the source
of your drinking water and how to preserve and protect it.

TechKNOWLEDGEy Strategic Group (http://tech-strategy.com): Colorado-
based independent investment banking and management consulting

238 RECOMMENDED RESOURCES

firm focusing on the commercial water and environmental services industry; founded by water expert and author Steve Maxwell.

Toxic Release Inventory Program (http://epa.gov/tri/tridata/tri03/index .htm): Get informed about the toxins and pollutants in your area.

UNESCO Water Portal (http://www.unesco.org/water/): The United Nations Educational Scientific and Cultural Organization's entry point to the current UNESCO and UNESCO-led programs on freshwater, and a resource for more information on other water-related organizations.

U.S. Army Corps of Engineers, Southeast Drought Response (http://www .sad.usace.army.mil/drought/conservationlinks.htm): Water conservation resources including Web links, fact sheets, and answers to frequently asked questions.

U.S. Drought Monitor (http://drought.unl.edu/dm/monitor.html): Quick link to the latest drought developments across the United States.

U.S. Environmental Protection Agency Office of Water (http://water.epa .gov/): A starting point to learn about your water and the latest water developments.

U.S. Geological Survey (http://www.usgs.gov): Government scientific research agency. Its web site provides independent information and reports on many topics, including:

    Current Streamflow Conditions (http://water.usgs.gov/waterwatch/?m= dryw&w=map&r=us): Check out the map to find out the latest streamflow conditions.

    Drip Accumulator (http://ga.water.usgs.gov/edu/sc4.html): Find out how that dripping faucet adds up to lots of water down the drain.

    Links to Other Water Information Sources (http://water.usgs.gov/ connections .html).

    Water Resources Information, WaterWatch (http://water.usgs.gov/ waterwatch): Shows maps, graphs, and tables describing real-time, recent, and past streamflow conditions for the United States.

    Water Science for Schools (http://ga.water.usgs.gov/edu/mwater .html): Great information on a variety of water and climate issues.

U.S. Global Change Research Program (http://globalchange.gov): Mandated by Congress as part of the Global Change Research Act of 1990, the program integrates federal research on climate and global change; known as the U.S. Climate Change Science Program, 2002 through 2008.

Water Environment Federation (http://wef.org): A longtime nonprofit that provides technical training and education to water quality professionals; heck out its "Water News" (http://wef.org/about/ IndustryNews.aspx) for some of the latest world water developments.

U.S. Water Monitor (http://watermonitor.gov): A portal to the federal water information on stream flows, reservoirs, groundwater, and forecasts.

Waterkeeper Alliance (http://waterkeeper.org): An international advocacy group dedicated to preserving and protecting water and waterways.

WateReuse Association (http://www.watereuse.org): A nonprofit with emphasis on water recycling and reuse; check out its National Database of Water Reuse Facilities, and community outreach web site (http://athirstyplanet.com).

World Bank (http://www.worldbank.org/water): Find out about global water issues.

ABOUT THE AUTHOR

Specializing in consumer issues, **Susan J. Marks** is an award-winning journalist with more than 30 years of experience. As a freelancer, she has written or collaborated on more than a dozen information, consumer, and self-help books, and has written for newspapers, magazines, and organizations across the country. She's helped dozens of people, organizations, governments, and agencies all over the world tell their stories.

Based in water-conscious Denver, Marks knows firsthand about water rights, restrictions, and shortages, and their repercussions. She has worked closely with Colorado ranchers and developers bent on expansion and curtailed by water rights and replenishment issues. She's also watched parched underbrush burn unchecked in water-starved areas of Colorado and Florida, and battled out-of-control brush fires on the other side of the world.

Prior to full-time freelancing, she spent more than a dozen years at *The Denver Post*, primarily as Sunday business editor and special projects editor for the business department. Her work has received awards and recognition from local, regional, and national professional and industry organizations.

INDEX